T0276104

THE PATH TO RESOLVE THE CMI MILLENIUM PROBLEMS

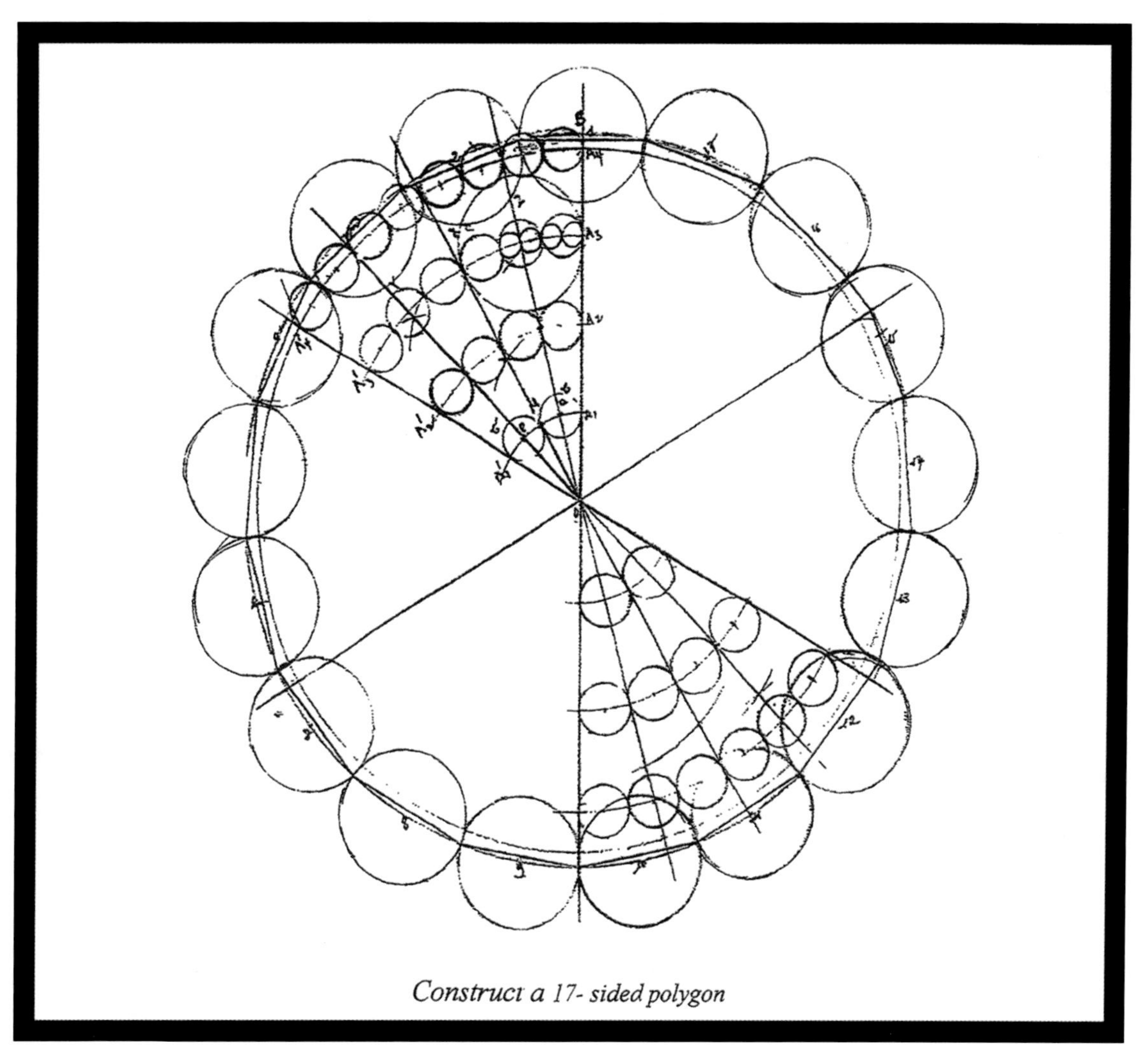

Construct a 17- sided polygon

Shi Feng Sheng
and Danny Wong

God created the universe and natural numbers, everything else is made by men. Infinite universe and natural numbers; macroscopic in distance, microscopic up close.

Galileo (1564-1642) once said:

* Nature is relentless and unchangeable, and it is indifferent as to whether its hidden reasons and actions are understandable to men or not.
* Mathematics is the language with which God has written the universe.
* We must say that there are as many squares as there are numbers.
* It is surely harmful to souls to make a heresy to believe what is proved.
* Where the sense fails us, reason must step in.
* All truths are easy to understand once they are discovered; the point is to discover them.
* Nothing occurs contrary to nature except the impossible, and that never occurs.

ACKNOWLEDGEMENT

THE PATH TO
RESOLVE
THE
CMI MILLENNIUM PROBLEMS

This Book
Would Not Be Possible
Without
The Valuable Insight Of A Book
Entitled
“The Millennium Problems”

By Keith Devlin

AuthorHouse™
1663 Liberty Drive
Bloomington, IN 47403
www.authorhouse.com
Phone: 833-262-8899

This book is printed on acid-free paper.

ISBN: 978-1-6655-5335-3 (sc)
ISBN: 978-1-6655-5336-0 (e)

Library of Congress Control Number: 2022903806

Print information available on the last page.

Published by AuthorHouse 06/15/2022

authorHOUSE®

Contents

PREFACE

[1] December 2009 was truly an extraordinary month in my life:

While dropping a friend off in downtown St Petersburg Florida; I met a distinguished looking elder Chinese (Mr. Shi). As we talked, somehow our conversation led him to claim that he had resolved the famous 1742 Goldbach's Conjecture (the principal unsolved problem in pure mathematics) originated from St Petersburg Russia as well as many other famous intractable problems in mathematics --- which I heard of during college years.

Both of us were amazed because: (a) it is very rare to see Asians in downtown St Petersburg, let alone to meet someone who speaking the same Shanghai dialogue, (b) Mr. Shi was presently surprised that I knew of his mathematical problems, and I was surprised of what he proclaimed, but skeptical. Nevertheless, we agreed to meet again soon.

Coincidently, my son (Timothy Wong), an amateur number theory enthusiast gave me a Math book entitled "The Millennium Problems" authored by Keith Devlin (Stanford) for Christmas a week earlier --- this book consists of seven intractable problems in the field of number theory, topology, physics and computer science selected by the "Clay Mathematical Institute". Moreover, CMI asks for their solutions with prize of $1.000.000 per/problem --- subject to certain constrains.

[2] When we met again the day after Christmas, Mr. Shi presented me with:

* A manuscript of his math book entitled "EXIST" --- which addressed two major unsolved problems in number theory namely; the 1742 Goldbach's conjecture, the 1637 Fermat's Last Theorem, and four classic Greek problems from (2000+ BC) --- "trisecting the angle", " heptagon", "doubling the cube" and "squaring the circle".
* Diagram of a (17-sided polygon) and claimed it was done by the first human (his high student) from Paris France --- via "Straightedge and Compass construction".

Undoubtedly, there will be serious ramification in the mathematics and beyond if the addressed topics in his "Book" and the "17-sided polygon" turned out to be credible.

[3] Months of tireless clarification of his texts (some were vague) in **EXIST** and the valuable in-sight of the subjects in "**The Millennium Problems**" --- renewed my curious spirit in mathematics --- to commit years of investigation, and concluded that the topics in **EXIST** were directly or indirectly linked to all the **CMI problems;** their connections are best understood to present them together in the (form) of this book because:

(a) With respect to his (six) topics addressed in EXIST: (i) papers on the **Goldbach's conjecture** and the **Fermat's Last Theorem** by famous mathematicians were already published by reputable Journals as a-step-in-the-right-direction because they were consistent with existing research activities, (ii) other **four problems of Antiquity** had declared by reputable mathematicians as "insoluble" a long time ago,

(iii) with respect to the **CMI problems**, they were based on modern abstract theories, theorems and conjectures --- that was (is) unintelligible even to the experts in the field.

Consequently, there are fair bits of upstream background to cover before I can begin to re-introduce the topics in EXIST, and lots of downstream abstractions to uncover before I can elaborate the CMI problems from our perspective.

(b) The (proclaimed) need to be studied together in the sequence we presented in order to illustrate that --- mathematical knowledge is not a collection of isolated facts. Each branch is a connected whole; linked to other branches that we do not understand mathematically, but ultimately, they are all connected to the roots of mathematics: **the pattern of the primes.**

(c) Due to the time consuming vetting rules stipulated by the CMI --- we decided that there is no time to waste because professor Shi is (87 years old); not mention that our proclaimed were totally inconsistent with the formal, the symbolic, the verbal, the analytic elements and modern abstract theories --- passed down by the famous or not-so-famous researchers. Nevertheless, this book is consistent with --- the nature mathematics (or arithmetic) passed down from the ancient Greeks and Fibonacci (1170-1240) and the famous quotes left by the great Galileo (1564-1642):

It is surely harmful to souls to make a heresy to believe what is proved.
Mathematics is the language with which God has written the universe.
We must say that there are as many squares as there are numbers.
Where the sense fails us, reason must step in.
Truths are easy to understand once they are discovered; the point is to discover them.
Nothing occurs contrary to nature except the impossible, and that never occurs.

More importantly, to illustrate the classic mechanics developed by Newton (1642-1726) was a precursor of the tremendous advancement in modern science and technology of yesterday and today.

Danny Wong
Sarasota Florida
January 2022
E-mail: existsfs@hotmail.com

ONE

Early History In Mathematics (BC to 19th century)

God created the universe and natural numbers, everything else is made by men. Infinite universe and natural numbers; macroscopic in distance, microscopic up close.

Natural number: 1, 2, 3, 4, 5, 6, 7, 8, 9, 10, 11. 13…
Even number: 2, 4, 6, 8, 10, 12, 14, 16, 18, 20...
Odd number: 1, 3, 5, 7, 9, 11, 13, 15, 17, 19…
Odd prime number: 3, 5, 7, 11, 13, 17, 19, 23, 29 …..

Prime numbers are any integers ≥ 2 that can be divided by 1 and itself only (two divisors), so 2 is the only even prime. Historically, there is no useful formula that yields all primes and no composites --- because they are not polynomial.

Historically, mathematicians have: (1) dealt with questions of finding and describing the intersection of algebraic curves, (2) wrestled with paradoxes of the "pattern of the prime numbers", "concept of infinite" and "sum of the infinite series" long before --- Euclid (300 BC) devoted part of his *Elements* to prime numbers and divisibility, topics the belong unambiguously belong to number theory; and introduced the 1st proof of infinitude of primes by abstract reasoning, the Euclidean geometry and considered line, circle as curves in geometry; the work of Archimedes (250 BC) and Sieve of Eratosthenes (240 BC).

In the 3rd century "Arithmetica" introduced Diophantine geometry --- a collection of problems giving numerical solutions of both determinate and indeterminate equations. Diophantine studied rational points on curves (elliptic) and algebraic varieties. In other words, Diophantine showed how to obtain infinitely many of the rational numbers satisfying a system of equations by giving a procedure that can be made into as an algebraic expression *(algebraic geometry in pure algebraic forms).* Diophantus contributed greatly in mathematical notation, and introduced approximate equality to find maxima for functions and tangent line to curves. Unfortunately, most of the texts were lost or unexplained.

Some of the roots of algebraic geometry date back to the work of Hellenistic Greeks (450 BC). The Delain problem --- "doubling the cube" and other related problems such as; "trisecting the angle", "polygons", "squaring the circle" --- they are also known as straightedge and compass problems (or topological problems).

In Geek mathematics --- *Number theory (or arithmetic)* is to study the "patterns of the numbers" and "elementary calculation technique", *geometry* is a technique to study patterns of shape, *algebra* is to study the patterns of putting things together, and *trigonometry* considers the measurement of shapes. *Topology* is to study the patterns of closeness and relative position. *Algebraic geometry* is to study geometry via algebraic curves and varieties. Not very much happened in mathematics after the Greeks until the 17th century:

* Galileo (1564-1642) was the father of "observational astronomy", "modern science" and a polymath in the field of mathematics, physics, engineering and natural philosophy.

In mathematics --- Galileo applied the standard (arithmetic) passed down from ancient Greeks and Fibonacci (1170-1240) but superseded later by the algebraic methods of Descartes.

* Rene' Descartes (1596-1630) made a fundamental discovery: Assuming by restrict ourselves to the "straightedge and compass" in geometry, it is impossible to construct segments of every length. If we begin with a segment of length 1, say, we can only construct a segment of another length if it can be expressed using integers, addition, subtraction, multiplication, division and square roots (as the golden ratio can). Thus, one strategy to prove that a geometric problem is unsolvable (not constructible) --- is to show that the length of some segment in the final figure cannot be written in this way. But doing so rigorously required the nascent field of algebra [...]

Descartes introduced the analytic geometry primarily to study algebraic curves to reformulate the (classic works on conic and cube); using Descartes' approach, the geometric and logical arguments favored by the ancient Greeks for solving geometric problems could be replaced by doing algebra *(algebraic geometry extended the mathematical objects to multidimensional and Non-Euclidean spaces).*

* Pierre de Fermat (1601-1665) independently developed the analytic geometry to study properties of algebraic curves (those defined in Diophantine geometry) which is the manifestation of solutions of a system of polynomial equations. Unfortunately, most of his work was in private letters or margin notes. *Fermat was famous for his (1637 FLT) and limited Fermat prime number --- 3, 5, 17, 257, 65537.*

* Desargues (1591-1661) introduced the projective geometry to study geometric properties that are invariant under projective transformations; it mainly dealt with those properties of geometric figures that are not altered by projecting their image onto another surface. Desargues also contributed in Blaise Pascal's Theorem.

Nevertheless, Desargues, Pascal (1623-1662) and others argued against the use of analytic and algebraic method to study geometry; they approached geometry from a different perspective, developing the synthetic notions of projective geometry they also studied curves, but from the purely geometrical point of view: the analog of the Greek ruler and compass construction.

In physics --- Galileo's theoretical and experimental work on the motions of bodies, along with the work of Kepler (1571-1630) and Descartes was a precursor of the classical mechanics developed by Newton (1642-1726).

Initially, calculus was introduced to study physical problems, capable of to approximate a polynomial series via manipulation. Historically, the first method of doing so was by infinitesimals, but infinitesimals do not satisfy the Archimedean property. Although calculus has widespread applications in science, engineering, economics and others, nonetheless, calculus is still to some extent an active area of research today.

* In 1715, Taylor series was introduced to represent a function as an infinite sum of terms that are calculated from the values of the function's derivatives at a single point. It became common practice to approximate a function by finite number of terms of its Taylor series.

Physicist/mathematician Daniel Bernoulli (1700-1782) accepted the 17th century analytic geometry because --- it supplied with concrete quantitative tools needed to analyze his physical fluid motion problem via infinitesimal calculus; his colleague Leonhard Euler (1707-1783) contributed three partial differential equations to describe Daniel Bernoulli's fluid flow and during the same period: (a) Euler became interested in number theory after his friend Christian Goldbach (1690-1764) introduced the work of Fermat (including the FLT) to him, (b) Euler also adapted infinitesimal calculus and his zeta function to investigate prime distribution:

$$\zeta(s) = \sum_{n=1}^{\infty} 1/n^s = 1 + 1/2^s + 1/3^s + 1/4^s + \ldots = \prod_{p}(1 - p^{-s})^{-1}$$

Although Euler was unsuccessful in all fronts but: (i) claimed his finite products by a **round-about** method, (ii) responsible for calculus to enter into natural mathematics for the first time, (iii) marked the "rebirth" of the 1637 (FLT) as the beginning of modern number theory.

In about 1742, after Euler announced the accuracy of (his friend) Goldbach's prime distribution observation but he could not prove it mathematically --- Goldbach's conjecture --- the principal problem in pure mathematics became a problem of analytic number theory. Consequently, Euler's analytic approaches influenced a new generation of researchers:

* The 1791 observation of prime distribution by a 14 years old Carl F. Gauss (1777-1855) in a sufficiently large N together with Legendre's (1752-1833) prime distribution conjecture became known as the "Gauss' density conjecture" --- due to their similarity
* In geometry, Gauss proved in theory that a "17-sided polygon" was constructible via "Straightedge and Compass construction" --- but no one was able to do so (including Gauss himself). In the beginning of 19th century; most of the algebraic characters was subsumed by infinitesimal calculus of Euler and Lagrange (1736-1813), and the mathematical community accepted the greatest puzzles in mathematics were:

Goldbach's conjecture (GC)

In about 1742, Christian Goldbach (1690-1764) conjectured that any even number $N \geq 6$ can be expressed as the sum of two odd prime numbers

Fermat's Last Theorem (FLT)

In about 1637, French mathematician Pierre de Fermat (1601- 1665) left a margin note in his copy of "Arithemtica" that this indeterminate Diophantine equation $A^N + B^N = C^N$ has no solution when N is an integer > 2. Furthermore, Fermat noted that he had a truly remarkable proof by (descent) but there was very little margin left to elaborate

Trisecting the angle (BC)

Polygons (BC)

Doubling the cube (BC)

Squaring the circle (BC)

(a) To put calculus in a more solid footing, Cauchy (1789-1857), a pioneer in early math analysis, defined continuity (formalized the concept of infinite).
(b) Wrierstrass (1815-1897) introduced elliptic curves in normal form as $Y^3 = x^2 + ax + b$, and gave the definition of limit (eliminated infinitesimal). Infinitesimals were replaced by very small numbers, and the infinitely small behavior of the function is found by taking the limiting behavior for smaller and smaller numbers. So, calculus is a collection of techniques for manipulating numbers and certain limits. Eventually, it became common to base calculus on limit and paved the way for modern study of number theory.
(c) Neil Abel (1802-1829) studied elliptic integrals, algebraic variety, complex plane and contributed a new class of Abelian functions and integrals [...]
(d) Jacobi (1804-1851) studied elliptic functions, partial derivative and others.
(e) L. Dirichlet (1805-1859) published "The existence of primes in a given arithmetic progression" in 1837, and his papers on "analytic number theory" in 1838 introduced his L-function associated with algebraic or arithmetical objects. Ty
(f) The 1849 observation of "Twin primes" by Alphonse de Polignac was officially coined by Paul Stackel (1862-1919) as the "Twin Primes conjecture".
(g) Riemann published his "Habilitation on the Foundation of Geometry" in 1953 and his famous "Riemann Hypothesis" in 1859.
(h) In 1896, Hadamard and de la Vallee Poussin independently published their "Prime Number Theorem" based on Gauss's density conjecture (but still with approximation).

(i) After centuries of exhaustive researches and abstractions:

* Although the (1637 FLT) and the (1742 GC) --- both dealt with positive integers (favor rigor), yet they were accepted as problems of analytic number theory of some fashion.
* Modern algebraic geometry is still in the uncharted area of mathematics, both conceptually and in terms of technique.
* In topology --- modern topologists do not mention names of those ancient Greek topological problems (or straightedge and compass problems) anymore --- because although they were geometric in nature, but the 17th century algebraic method of Descartes influenced:

(a) The 1837 paper by Pierre Wantzel (1814-1848) to use polynomial (the sums of coefficients and variables raised to powers) and their roots (values that make the polynomial equal to zero) [....] to prove the "Doubling the cube" was "unsolvable", moreover, he dismissed ---"trisecting the angle" and "polygons" as unsolvable via the same reasons --- unfortunately, his paper was discovered in the late 19th century.
(b) "Squaring the circle" was proved by Ferdinand von Lindermann (1852-1939) as "unsolvable" in his 1882 paper --- claiming that π was not constructible [...]

It is worth noting --- Dirichlet (1805-1859) was Riemann's teacher in U of Berlin, and Gauss (1777-1855) was Riemann's PhD advisor in U of Gottingen, and they chaired the Dept. of Mathematics in Gottingen consecutively (The 19th century authorities of Mathematics and Science). Historically, it was the words of authority that matters.

TWO

Recent History In Mathematics (20th to 21st century)

* In 1900, sequence of events in early history of mathematics and the 17th century: (a) analytic geometry, (b) the elusive 1637 (FLT), (c) calculus of Newton, along with the analytic elements of the 18th and 19th centuries --- influenced David Hilbert to announce his 23 important mathematical puzzles at Paris inaugural International Math Conference.

In the field of number theory, Hilbert listed "Riemann Hypothesis, Goldbach's conjecture, Twin Primes conjectures" and "The solvability of a Diophantine equation" as his 8th and 10th problems; and subsequently, modern abstract theories, theorems and conjectures of rational points on the elliptic curves prospered in the early 20th century, and consequently, mathematicians believed that --- problems in number theory are best understood through study of analytical objects (e.g., Riemann's zeta function) that encodes properties of the integers, or other number-theoretic objects in some fashion (analytic number theory).

* In 1990, an advanced mathematical book entitled "EXIST" authored by an unknown Chinese Naval architect/engineer. The book was copyrighted with the Library of Congress. Registration Number & Date: TXU0003465073/1990 – 12 – 4.
Contents: Goldbach's Conjecture, an infinite Calculation Chart, Fermat's Last Theorem,
Trisecting the angle, heptagon, doubling the cube, squaring the Circle

This book was never published due to Mr. Shi's personal reasons; but please refers to **(Appendix A)** in CHAPTER FOUR and notices that:

(a) The Goldbach's conjecture was verified by Arthur Wightman (Princeton) in 1989.
(b) The Chart was acknowledged by The National Geographic Society in 1994.
(c) The book was acknowledged by Pre Bush, VP Gore and other Institutions.

* In 2000, CMI announced their millennium prize problems as:

EXHIBIT 1: P = NP in P vs. NP (1971) ------------------------------------- a major computer science problem
EXHIBIT 1-A: P ≠ NP in P vs. NP (1971)---------------------------------- a major computer science problem
EXHIBIT 2: Riemann Hypothesis (1859) ------------------------------------ refers to the pattern of the primes
EXHIBIT 3: Birch & S-D conjecture (1960) ------------------------ modern study of a Diophantine equation
EXHIBIT 4: Poincare conjecture (1905) -- a modern study of topology
EXHIBIT 5: Hodge conjecture (1950) --- a modern study of topology
EXHIBIT 6: The Yang-Mills Theory (1954) --------------------------------------a theory in quantum physics
EXHIBIT 7: Naiver–Stokes Equations (1820's) -------------------------------an important problem in physics

Apparently:

(a) All the CMI problems required a rigorous proof to infinity (to mean polynomial). Obviously, the existing knowledge has failed --- that is why they have a price tag of $1,000,000.

(b) The P vs. NP problem is the only computer science problem in CMI grew out of "The complexity of theorem proving procedures" introduced by Stephen Cook in 1971; other CMI problems grew out of --- the formal, the symbolic, the verbal, the analytic elements and modern abstract theories passed down by the famous or not-so-famous researchers.

* In 2002, at the request of CMI, Keith Devlin (Stanford) published his book entitled "The Millennium Problems" to introduce the CMI problems in a more transparent fashion. In which, Devlin stated "For a century now, mathematicians have built new abstractions on top of the old ones, every new step taking them further from the world of everyday experience on which, ultimately, we must base all our standing. It is not so much that mathematician does new things; rather, the object considered became more abstract, abstractions from abstractions from abstractions.

Indeed, all the CMI problems were direct or indirectly related to the topics addressed in an **unpublished** book:

Title: EXIST/ Sheng, Shi Feng.
Copyrighted with the Library of Congress
Registration Number & Date: TXU000465073/1990-12-04.

Contents:
EXHIBIT A: Goldbach's conjecture (1742) ------------------------------- refers to the pattern of the primes
EXHIBIT B: An infinite calculation chart and reading instruction
EXHIBIT C: Fermat's Last Theorem (1637) ------------------------ an old Diophantine geometry problem
EXHIBIT D: Trisecting the angle ------------------------------- to divide any given angle into 3 equal parts
EXHIBIT E: Heptagon --------------------------------- to construct a polygon of any given number of sides
EXHIBIT F: Doubling the Cube -----------------------to construct one side of a cube to double its volume
EXHIBITG: Squaring the Circle ---------------------- to draw a square has the same area of a given circle

These topics illustrated --- mathematical knowledge is not a collection of isolated facts. Each branch is a connected whole; linked to other branches that we could not understand mathematically, but ultimately, they are all connected to the roots of mathematics: the pattern of the primes.

****

In essence, we are claiming **(Exhibit A to G)** addressed in EXIST were the essential ingredients needed to resolve the CMI problems in **(Exhibit 1 to** 7). More importantly, the sequence of their connections is MAPPED OUT in next page.

****

===

Exhibit A: A proof of the Goldbach's conjecture

===

↓

==

Exhibit B: An infinite calculation chart mapped out based on proving the Goldbach's principle (1+1) that outperforms any computer in calculation, both in speed and range. It also represents as an infinite topological field that led to resolve those unsolvable problems of Antiquity --- "Trisecting the angle", "Heptagon", "Doubling the cube", and "squaring the circle" ; please refer to (Exhibit D, E, F and G):

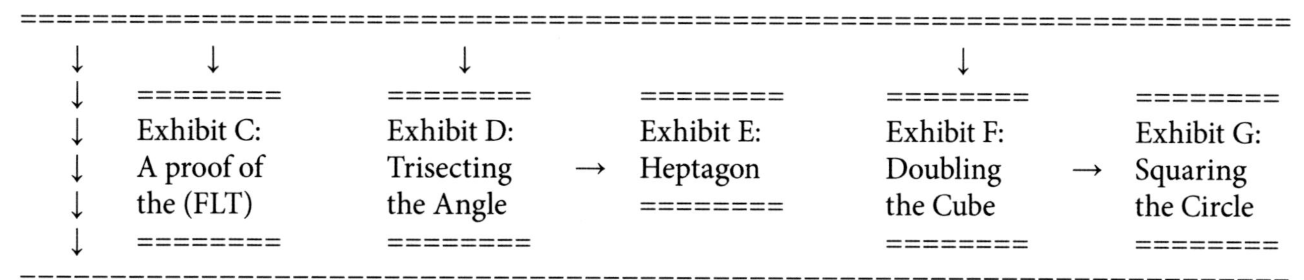

==

(CMI) Exhibit 1: A proof of the P = NP in P vs. NP. Author Stephen Cook implied that if P = NP is proven, then it may lead: (a) to resolve a certain calculation tasks that a computer fails to perform, (b) transform mathematics by allowing a computer to find a formal of any theorem which has a proof of a reasonable length since formal proof can be recognized in polynomial time. Example problems may include all the CMI problems

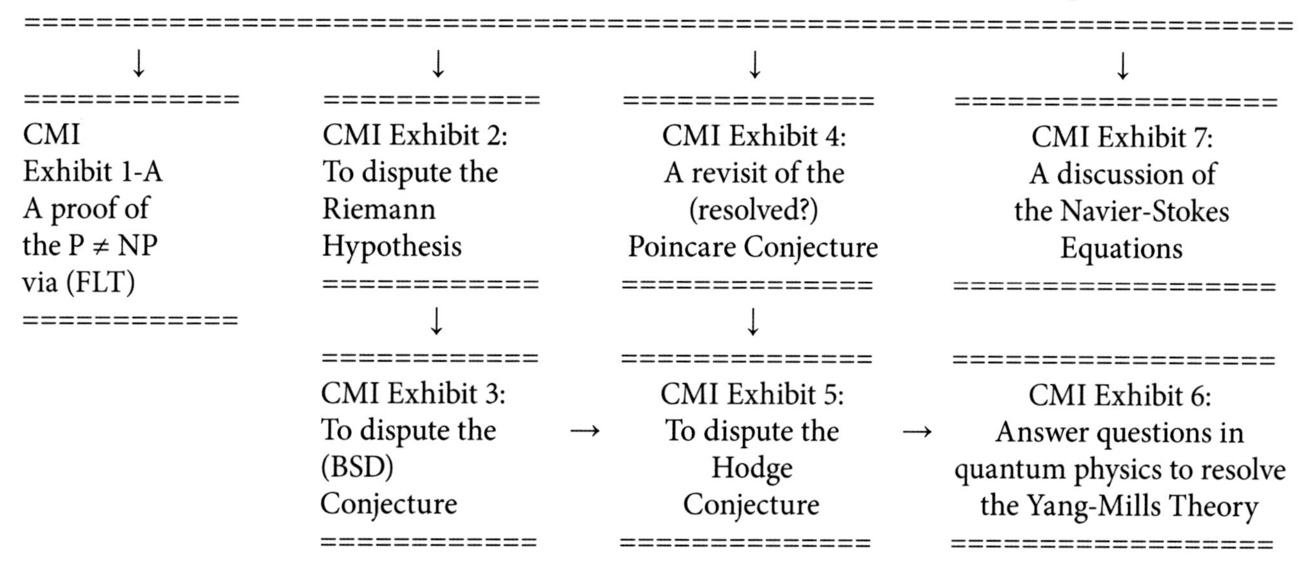

Mathematicians have no perception of the texts in **(Exhibit A to G),** both conceptually and in terms of technique; and the CMI problems in **(Exhibit 1 to 7)** were presented in a (FORMAT) --- so we don't have to repeat the same lengthy topics from (EXIST) as --- REFERENCE, CROSS REFERENCE or COUNTEREXAMPLE.

It is imperative for researchers in the field to review these (15 EXHIBITS) together with an open mind --- since the proclaimed were totally inconsistent with existing knowledge and other standard that you are accustomed to.

EXHIBIT A A PROOF OF THE GOLDBACH'S CONJECTURE (GC)

ABSTRACT. Contrary to existing wisdom that problems in number theory are best understood through study of analytical objects (e.g., Riemann's zeta function) that encodes properties of the integers, or other number-theoretic objects in some fashion (analytic number theory); our objective is to illustrate --- an overlooked 18th century technique we met, along with other logical observation was a precursor of: (a) an uncanny polynomial procedure to explain where every odd primes will occur in the range of any N, (b) an unprecedented polynomial algorithm (of a particular form) to hold the (GC) true infinitely via studying the "patterns of the whole numbers" and "reasoning".

1. THE PROBLEM

In 1742, Christian Goldbach (1690-1764) observed prime distribution as: (a) any even number $N \geq 6$ can be expressed as the sum of two odd prime numbers. After his friend Leonhard Euler (1707-1783) verified its accuracy but could not solve it mathematically, (GC) became the principal problem in pure mathematics, (b) a weak (GC) is also known as any odd number ≥ 9 can be expressed as the sum of three odd primes.

2. OBSERVATION AND REASONING

(1) Prior to the 3rd century "Arithmetica", Euler used zeta function and infinitesimal calculus to investigate prime distribution in the 1735(s) that marked the "rebirth" of the elusive 1637 (FLT) as the beginning of modern number theory; number theory (or arithmetic) was to study the "patterns of the numbers" and "elementary calculation techniques".

Prior to the 1791 observation of prime distribution by Carl F. Gauss's (1777-1855) together with Legendre's (1752-1833) prime conjecture --- were accepted as Gauss' density conjecture in the 19th century; it was well known that Gauss studied the patterns of the numbers when he was 7 years old, and found a short cut to the solution of the sum of first 100 numbers using the technique of ascending and descending order of nature numbers:

$$\begin{array}{l} 1 + 2 + 3 + 4 + \dots + 97 + 98 + 99 + 100 \\ 100 + 99 + 98 + 97 + \dots + 4 + 3 + 2 + 1 \end{array}$$

and reasoned that each vertical pair adds up to 101; there are 100 pair vertical pairs. Sum of all the numbers in both lines is (100 x 101= 10100). But this will be exactly twice the sum of the numbers in either of the two rows. So, the sum of the numbers in either two rows will be (½ of 10100 = 5050). In fact, Gauss's short cut is equivalent of ascending and descending order of the first nine numbers in the range of N = 10:

$$\begin{array}{l} 1 + 2 + 3 + 4 + 5 + 6 + 7 + 8 + 9 \\ 9 + 8 + 7 + 6 + 5 + 4 + 3 + 2 + 1 \end{array}$$

Each vertical pair adds up to 10, and there are 9 pairs, therefore, ($9 \times 10 = 90$). But this will be twice the sum of the numbers in either two rows. So the sum of the numbers in either two rows will be (½ of 90 = 45).

Let N = 9, then N (N + 1) / 2 = (9 × 10) / 2 = 90 / 2 = 45. In fact, N (N+1) / 2 are a polynomial equation (of a particular form) to calculate partial sum of the infinite series of 1 + 2 + 3 + ...+ N (in a reasonable range):

	N (N+1) / 2 = 1 + 2 + 3 +...+ N Let: N = any positive integer
N = 9:	N (N+1) / 2 = 9 (10) / 2 = 45
N = 10:	N (N+1) / 2 = 10 (11) / 2 = 110 / 2 = 505
$N = 10^2$	N (N+1) / 2 = 100 (101) / 2 = 10010 / 2 = 5050
$N = 10^{100}$	It will take exponential time that made impossible to calculate

In fact, any problems in number theory that deals with positive integers can be studied via the patterns of their numbers in the range of N = 10 --- because positive integers has an arithmetical progression of 1, 10, 100, 1000, 10000 ... with exception of "1"; all the other numbers are based on 1 + 1, so "9" is the largest number due to 10 is a new"1"; so that $11 = 1.1 \times 10$, $101 = 1.01 \times 10^2$, $1001 = 1.001 \times 10^3$ ...

(2) Goldbach's conjecture deals with positive integers, favor rigor (to mean polynomial).

(GC) is also known as (1+1) --- to mean all the even number > 2 can be expressed as the sum of two odd primes. Naturally --- 2, 3, 5, 7 are primes in the range of N = 10, so (4 = 2 + 2), (6 = 3 + 3), (8 = 3 + 5), (10 = 3 + 7), (10 = 5 + 5) complied with (GC); and 4, 6, 8 ... are composites with a common prime factor 2, and naturally, prime 2 is the building block of the odd prime and composite numbers. For example:

1 + 2 = **<u>3</u>**	3 + 2 = **<u>5</u>**	5 + 2 = **<u>7</u>**	7 + 2 = 9	9 + 2 = **<u>11</u>**	11 + 2 = **<u>13</u>**
13 + 2 = 15	15 + 2 = **<u>17</u>**	17 + 2 = **<u>19</u>**	19 + 2 = 21	So on to the forth	

Moreover, the pattern 3, 5, 7, 9, 11, 13, 15, 17 ... coupled with (Sieve theory) inspired an uncanny polynomial procedure to explain where every odd primes will occur in the range of any given N. For details, please refer to **(Appendix A-1)** in CHAPTER FOUR.

(3) Odd primes exist in odd numbers only, ascending and descending order of the odd numbers in the range of N = 10:

1 + 3 + 5 + 7 + 9 = 25
9 + 7 + 5 + 3 + 1 = 25

automatically showed two vertical pair (10 = 3 + 7), (10 = 5 + 5) that comply with the conjecture; a new idea is discovered to study the (GC) via ascending and descending order of the odd numbers **(VERTICALLY)** in the increment of any $N \geq 6$. For example:

N = 10: The upper half has: three primes --- 3, 5 and two odd numbers --- 1, 9

1 + 9	The pattern of the odd numbers in the range N = 10 naturally yield two
3 + 7 X	pairs of (X) that comply with the (GC) --- (10 = 3 + 7). (10 = 5 + 5)
5 + 5 X	
-------	Since prime 5 is a (repeated number), therefore, the (upper) and (lower)
7 + 3	half have the same three primes (3, 5. 7) and two odd numbers (1, 9)
9 + 1	

N = 18:

1 + 17
3 + 15
5 + 13 X
7 + 11 X
9 + 9
----
11 + 7
13 + 5
15 + 3
17 + 1

The upper half has:
6 primes --- 3, 5, 7, 11, 13, 17
3 odd numbers --- 1, 9, 15
Obtained 2 pairs: (18 = 5 + 13), (18 = 7 + 11) comply with (GC)
1 pair of odd numbers (18 = 9 + 9)

Notice (18 = 9 + 9) has (repeated odd number 9).

N = 22:

1 + 21
3 + 19 X
5 + 17 X
7 + 15
9 + 13
11 + 11 X
----
13 + 9
15 + 7
17 + 5
19 + 3
21 + 1

The upper half has:
7 primes --- 3, 5, 7, 11, 13, 17, 19
4 odd numbers --- 1, 9, 15, 21
Obtained 3 pairs in the range of N = 22 that comply with (GC),
(22 = 3 + 19)
(22 = 5 + 17)
(22 = 11 + 11) (prime 11 is a repeated number).

Moreover, obtained (22 = 1 + 21) --- one pair of (odd number) in the range of N = 22.

N = 32:

1 + 31
3 + 29 X
5 + 27
7 + 25
9 + 23
13 + 19 X
11 + 21
15 + 17
----
17 + 15
19 + 13
21 + 11
23 + 9
25 + 7
27 + 5
29 + 3
31 + 1

The upper half has:
10 primes --- 3, 5, 7, 11, 13, 17, 19, 23, 29, 31
6 odd numbers --- 1, 9, 15, 21, 25, 27
Obtained 2 pairs of primes in the range of N = 32 that comply with the (GC) --- (32 = 3 + 29), (32 = 13 + 17)

Notice that:

* Odd primes exist in odd numbers in the range of any even even number N ≥ 6.

* Nature pattern of the odd numbers in the range of N = 32 showed there is (zero pair of odd numbers)

* (Total numbers in the upper half) – (Total numbers in the lower half) = 0

N = 42:

1 + 41	The upper half has:
3 + 39	10 primes --- 3, 5, 7, 11, 13, 17, 19, 23, 29, 31, 37, 41
5 + 37 X	6 odd numbers --- 1, 9, 15, 21, 25, 27, 33, 35, 39
7 + 35	Obtained 4 pairs comply with (GC):
9 + 33	(42 = 5 + 37)
11 + 31 X	(42 = 11 + 31)
13 + 29 X	(42 = 13 + 29)
15 + 27	(42 = 19 + 23)
17 + 25	
19 + 23 X	Obtained (42 = 9 + 33), (42 = 15 + 27), (42 = 21 + 21) ---
21 + 21	(3 pairs of odd numbers) in the range of N = 42 (but one
-----------	pair has repeated odd number 21).
23 + 19	
↓	

When **N = 100:**

The same technique showed 5 pairs that comply with the (GC) --- (100 = 3 + 97), (100 = 11 + 89), (100 = 17 + 83), (100 = 29 + 71), (100 = 47 + 53).

Accordingly, after randomly tested 10 even numbers in the range of N = 200, we obtained every pairs that complied with the (GC)

It seems we have ample empirical evidence to say that the (GC) is correct, but it is impossible to hold Goldbach's statement true infinity --- since in mathematics, "infinite" is greater in value than any specified number, however large.

4. THE GENERAL FORMULA

Here, a general formula (of a particular form) was uncovered through an unprecedented polynomial procedure to give exact number pair of odd primes to comply with the conjecture in the range of any $N \geq 6$. Using plain mathematical language; let X represents the number of pairs of primes that satisfy the Goldbach's statement within N. So there exists one pair if X = 1, and none if X = 0. In case that X > 1, the result is even more promising because we can find more than one pair. This certainty is conveyed in the following logical deduction:

The formula is $X = K + p / 2 - d / 2$

N: any even number ≥ 6
p: Number of prime numbers
d: Number of odd numbers (other than p)
K: Number of pairs of d within N, the sum of each pair being equal to N
X: Number of pairs of p within N, the sum of each pair being equal to N

$$X = K + p / 2 - d / 2$$

Variation: $d = 2K + p - 2X$

Assuming X = 0, then: $d = 2K + p$

When p (# of primes) becomes rare as N increases, d (# of odd numbers) grows close accordingly. This tendency means that K will increase, and consequently, this requires that X increase too in the context of above formula. Therefore, X must have a value. They are illustrated unambiguously in the following example:

N = 6: The upper half has:

p = 2, d = 1

1 + 5 K = 0

3 + 3X X = K + p / 2 - d / 2

------ = 0 + 2 /2 – 1 /2

5 + 1 = ½ (1/2 pair because the repeated prime 3, so (6 = 3 + 3)

N = 10:

The upper half has:

1 + 9 K p = 3 primes, d = 2 odd numbers, K = 1

3 + 7 X X = K + p / 2 – d / 2

5 + 5 X = 1 + (3 – 2) / 2

------- = 1 ½ (1/2 pair because of the repeated prime 5)

7 + 3 Obtained: (10 = 3 + 7), (10 = 5 + 5)

9 + 1

Let: N = 30:

p: Number of prime numbers
d: Number of odd numbers (other than p)
K: Number of pairs of d within N, the sum of each pair being equal to 30
X: Number of pairs of p within N, the sum of each pair being equal to 30

1 + 29 The upper half has:

3 + 27 p = 9

5 + 25 d = 6

X 7 + 23 K = 1 ½ (1/2 because of the repeated odd number 15)

9 + 21K

X11 + 19 X = K + p / 2 – d / 2

X13 + 17 = 1 ½ + (9 – 6) / 2

15 + 15K = 3

----

17 + 13 Obtained 3 pairs:

19 + 11

↓ (30 = 7 + 23) (30 = 11 + 19), (30 = 13 + 17)

Let: N = 40:

p: Number of prime numbers

d: Number of odd numbers (other than p)

K: Number of pairs of d within N, the sum of each pair being equal to 40

X: Number of pairs of p within N, the sum of each pair being equal to 40

1 + 39	The upper half has:
X 3 + 37	p = 11
5 + 35K	d = 9
7 + 33	K = 2
9 + 31	
X11 + 29	X = K + p / 2 – d / 2
13 + 27	= 2 + (11 – 9) / 2
15 + 25K	= 3
X17 + 23	
19 + 21	
----------	Obtained 3 pairs:
21 + 19	(40 = 3 + 37) (40 = 11 + 29), (40 = 17 + 23)
↓	

Let: N = 50:

P: Number of prime numbers

d: Number of odd numbers (other than p)

K: Number of pairs of d within N, the sum of each pair being equal to 50

X: Number of pairs of p within N, the sum of each pair being equal to 50

1 + 49 K	The upper half has:
X 3 + 47	p = 14 (3,5,7,11,13,17,19, 23, 29, 31, 37, 41,43, 47) .
5 + 45	d = 11 (1, 9, 15, 21, 25, 27, 33, 35, 39,45, 49)
X 7 + 43	K = 3½ (half because the repeated odd number 25)
9 + 41	
11 + 39	X = K + p / 2 – d / 2
X13 + 37	= 2½ + (14 – 11) / 2
15 + 35 K	= 2½ + 1½
17 + 33	= 4
X19 + 31	
21 + 29	Obtained 4 pairs:
23 + 27	
25 + 25 ½ K	(50 = 3 + 37), (50 = 7 + 43)
-----------	50 = 13 + 37), (50 = 19 + 31)
↓	

Significance of the above examples is that: (i) each solution of the (GC) is independent, (ii) solution time is proportional to the N involved, (iii) after randomly tested 10 even numbers N by brute force, it became clear that this general formula will hold the (GC) true at any even number N (in a reasonable range). For example:

$N = 116$: $p = 29$ $d = 29$ $K = 6$

$X = K + p / 2 - d / 2 = 6 + (29 / 2 - 29 / 2) = 6$

(1) $116 = 3 + 113$
(2) $116 = 7 + 109$
(3) $116 = 13 + 103$
(4) $116 = 19 + 97$
(5) $116 = 37 + 79$
(6) $116 = 43 + 73$

$N = 120$: $p = 29$ $d = 31$ $K = 13$

$X = K + p / 2 - d / 2 = 13 + (29 - 31) / 2 = 12$

(1) $120 = 7 + 113$
(2) $120 = 11 + 109$
(3) $120 = 13 + 107$
(4) $120 = 17 + 103$
(5) $120 = 19 + 101$
(6) $120 = 23 + 97$
(7) $120 = 31 + 89$
(8) $120 = 37 + 83$
(9) $120 = 41 + 79$
(10) $120 = 47 + 73$
(11) $120 = 53 + 67$
(12) $120 = 59 + 61$

$N = 400$: $p = 77$ $d = 123$ $K = 37$

$X = K + p / 2 - d / 2 = 37 + (77 / 2 - 123 / 2) = 14$

$N = 500$: $p = 94$ $d = 156$ $K = 44$

$X = K + p / 2 - d / 2 = 44 + (94 / 2 - 156 / 2) = 13$

(1) $500 = 13 + 487$
(2) $500 = 37 + 463$
(3) $500 = 43 + 457$
(4) $500 = 61 + 439$
(5) $500 = 67 + 433$
(6) $500 = 79 + 421$
(7) $500 = 103 + 397$
(8) $500 = 127 + 373$
(9) $500 = 151 + 349$
(10) $500 = 163 + 337$
(11) $500 = 193 + 307$
(12) $500 = 223 + 277$
(13) $500 = 229 + 271$

$N = 900$: $p = 153$ $d = 297$ $K = 120$

$X = K + p / 2 - d / 2 = 120 + (153 / 2 - 297 / 2) = 48$

(1) $900 = 13 + 887$
(2) $900 = 17 + 883$
(3) $900 = 19 + 881$
(4) $900 = 23 + 877$
(5) $900 = 37 + 863$
(6) $900 = 41 + 859$
(7) $900 = 43 + 857$
(8) $900 = 47 + 853$
(9) $900 = 61 + 839$
(10) $900 = 71 + 829$
(11) $900 = 73 + 827$
(12) $900 = 79 + 821$
(13) $900 = 89 + 811$
(14) $900 =103 + 797$
(15) $900 = 113 + 787$
(16) $900 = 127 + 773$
(17) $900 =131 + 769$
(18) $900 = 139 + 761$
(19) $900 = 149 + 751$
(20) $900 =157 + 743$
(21) $900 = 167 + 733$
(22) $900 = 173 + 727$
(23) $900 =181 + 719$
(24) $900 = 191 + 709$
(25) $900 = 199 + 701$
(26) $900 =223 + 677$
(27) $900 = 227 + 673$
(28) $900 = 239 + 661$
(29) $900 =241 + 659$
(30) $900 = 257 + 643$
(31) $900 = 269 + 631$
(32) $900 =281 + 619$
(33) $900 = 283 + 617$
(34) $900 = 293 + 607$
(35) $900 = 307 + 593$
(36) $900 = 313 + 587$
(37) $900 = 331 + 569$
(38) $900 = 337 + 563$
(39) $900 = 353 + 547$
(40) $900 = 359 + 541$
(41) $900 = 379 + 521$
(42) $900 = 397 + 503$

(43) $900 = 401 + 499$
(44) $900 = 409 + 491$
(45) $900 = 421 + 479$
(46) $900 = 433 + 467$
(47) $900 = 439 + 461$
(48) $900 = 443 + 457$

$N = 1000$: $p = 167$ $d = 333$ $K = 110$

$X = K + p / 2 - d / 2 = 110 + (167 / 2 - 333 / 2) = 27$

N = 5000:   p = 668   d = 2332   K = 945   X = 113

X = K + p / 2 – d / 2 = 945 + (668 / 2 – 2332 / 2) = 113

5. THE CERTAINTY

Now, let's address three particular [situations] to elucidate the certainty that Goldbach's statement will hold true infinitely --- via reasoning:

[Situation 1]: N = 116,   p = 29,   d = 29,   K = 6   Notice: (p = d = 29)

p: Number of prime numbers in the range of N = 116
d: Number of odd numbers (other than p)
K: Number of pairs of d within N, the sum of each pair being equal to 116
X: Number of pairs of p within N, the sum of each pair being equal to 116

1 + 115 K
X 3 + 113
5 + 111
X 7 + 109
9 + 107
11 + 105
X13 + 103
15 + 101
17 + 99
X19 + 97
21 + 95 K
23 + 93
25 + 91 K
27 + 89
29 + 87
31 + 85
33 + 83
35 + 81 K
X37 + 79
39 + 77 K
41 + 75
X43 + 73
45 + 71
47 + 69
49 + 67
51 + 65 K
53 + 63
55 + 61
57 + 59
----
59 + 57
↓

In upper half, we have:

p = 29   (3, 5, 7, 11, 13, 17, 19, 23, 29, 31, 37, 41, 41, 43, 47, 53, 59, 61, 67, 71, 73, 79, 83, 89, 97, 101, 103, 107, 109, 113)

d = 29   (1, 9, 15, 21, 25, 27, 33, 35, 39, 45, 49, 51, 55. 57, 63, 65, 69, 75, 77, 81, 85, 87, 91, 95, 99, 101, 105, 111, 115)

K = 6

X = K + p / 2 – d / 2
 = 6 + 29 / 2 – 29 / 2
 = 6

(1) 116 = 3 + 113   (2) 116 = 7 + 109
(3) 116 = 13 + 103   (4) 116 = 19 + 97
(5) 116 = 37 + 79   (6) 116 = 43 + 73

When: N = 120 p = 29 d = 31 K = 13 X = 12

p: Number of prime numbers in the range of N = 120
d: Number of odd numbers (other than p)
K: Number of pairs of d within N, the sum of each pair being equal to 120
X: Number of pairs of p within N, the sum of each pair being equal to 120

N = 120

1 + 119
3 + 117
5 + 115
X 7 + 113
9 + 111
X 11 + 109
X 13 + 107
15 + 105 K
X 17 + 103
X 19 + 101
21 + 99 K
X 23 + 97
25 + 95 K
27 + 93 K
29 + 91
X 31 + 89
33 + 87 K
35 + 85 K
X 37 + 83
39 + 81 K
X 41 + 79
43 + 77
45 + 75 K
X 47 + 73
49 + 71
51 + 69 K
X 53 + 67
55 + 65 K
57 + 63 K
X 59 + 61

↓

The upper half has:

p = 29 (3, 5, 7, 11, 13, 17, 19, 23, 29, 31, 37, 41, 43, 47, 53, 59, 61, 67, 71, 73, 79, 83, 89, 97, 101, 103, 107, 109, 113)
d = 31 odd numbers (1, 9, 15, 21, 25, 27, 33, 35, 39, 45, 51, 53, 55, 57, 63, 65, 69, 75, 77, 81, 85, 87, 91, 93, 95, 99, 105, 111, 115, 117, 119)
K = 13

X = K + p / 2 – d / 2
= 13 + 29 / 2 – 31/ 2
= 13 – 1
= 12 (pairs that complied with GC)

(1) 120 = 7 + 113
(2) 120 = 11 + 109
(3) 120 = 13 + 107
(4) 120 = 17 + 103
(5) 120 = 19 + 101
(6) 120 = 23 + 97
(7) 120 = 31 + 89
(8) 120 = 37 + 83
(9) 120 = 41 + 79
(10) 120 = 47 + 73
(11) 120 = 53 + 67
(12) 120 = 59 + 61

Please notice the change in these two cases:
N: 116 ⟶ 120, p: both = 29, d: 29 ⟶ 31, K: 6 ⟶ 13
The fact is that when N = 116, K = 6. When N = 120, d has increased by 2 and K has increased by 7. We can only infer that K increases in pace with the growth of N value.

[Situation 2]: $N = 900$ $p = 153$ $d = 297$ $K = 120$ $X = 48$

* Let us divide N into two equal parts (450/450). For the first half, let p_1 and d_1 represent the number of primes and the number of odd numbers respectively. So are p_2 and d_2 for the second half. We obtain:

$p_1 = 86$ $p_2 = 67$
$d_1 = 139$ $d_2 = 158$

In the context of: $X = K + p / 2 - d / 2$
Variation: $2X = 2K + p - d$
$2X - 2K = p - d$
Assuming $X = 0$: $2K = d - p$
$K = (d - p) / 2$

This means each p_1 corresponds to one d_2, so is the relation between d_1 and p_2. So K becomes:

$d_2 - p_1 = 158 - 86 = 72$
$d_1 - p_2 = 139 - 67 = 72$

It seems impossible to corresponding one d_2 with p_1, or one p_2 with one d_1. In fact, the chances become rare or even hopeless as N increases. Nevertheless, the growing nature of K value demands that X should increase appropriately. Notice that we talk about "increase" in reference to the situation N = 116, K = 6 and X= 6 **(see page 14)**; in other words, we explore the increment of X on the assumption that (GC) is correct.

(1) 900 = 1 + 899
(2) 900 = 9 + 891
(3) 900 = 15 + 885
(4) 900 = 21 + 879
(5) 900 = 25 + 875
(6) 900 = 27 + 873
(7) 900 = 33 + 887
(8) 900 = 35 + 885
(9) 900 = 39 + 861
(10) 900 = 45 + 855
(11) 900 = 49 + 851
(12) 900 = 51 + 849
(13) 900 = 55 + 845
(14) 900 = 57 + 843
(15) 900 = 63 + 837
(16) 900 = 65 + 835
(17) 900 = 69 + 831
(18) 900 = 75 + 825
(19) 900 = 81 + 819
(41) 900 = 159 + 741
(42) 900 = 165 + 735
(43) 900 = 169 + 731
(44) 900 = 171 + 729
(45) 900 = 175 + 725
(46) 900 = 177 + 723
(47) 900 = 183 + 717
(48) 900 = 185 + 715
(49) 900 = 187 + 713
(50) 900 = 189 + 711
(51) 900 = 195 + 705
(52) 900 = 201 + 699
(53) 900 = 203 + 697
(54) 900 = 205 + 695
(55) 900 = 207 + 693
(56) 900 = 213 + 687
(57) 900 = 215 + 685
(58) 900 = 219 + 681
(59) 900 = 221 + 679
(81) 900 = 309 + 591
(82) 900 = 315 + 585
(83) 900 = 319 + 581
(84) 900 = 321 + 579
(85) 900 = 325 + 575
(86) 900 = 327 + 573
(87) 900 = 333 + 567
(88) 900 = 335 + 565
(89) 900 = 359 + 561
(90) 900 = 341 + 559
(91) 900 = 345 + 555
(92) 900 = 351 + 549
(93) 900 = 355 + 545
(94) 900 = 357 + 543
(95) 900 = 361 + 539
(96) 900 = 363 + 537
(97) 900 = 365 + 535
(98) 900 = 369 + 531
(99) 900 = 371 + 529

(20) 900 = 85 + 815
(21) 900 = 87 + 813
(22) 900 = 93 + 807
(23) 900 = 95 + 805
(24) 900 = 99 + 801
(25) 900 =105+ 795
(26) 900 =111+ 789

(27) 900 =115+ 785
(28) 900 =117+ 783
(29) 900 =119+ 781
(30) 900 =121+ 779
(31) 900 =123+ 777
(32) 900 =125+ 775
(33) 900 =129+ 771
(34) 900 =133+ 767
(35) 900 =135+ 765
(36) 900 =141+ 759
(37) 900 =145+ 755
(38) 900 =147+ 753
(39) 900 =153+ 747
(40) 900 =155+ 745

(60) 900 = 225 + 675
(61) 900 = 231 + 669
(62) 900 = 235 + 665
(63) 900 = 237 + 663
(64) 900 = 243 + 657
(65) 900 = 245 + 655
(66) 900 = 249 + 651

(67) 900 = 255 + 645
(68) 900 = 261 + 639
(69) 900 = 265 + 635
(70) 900 = 267 + 633
(71) 900 = 273 + 627
(72) 900 = 275 + 625
(73) 900 = 279 + 621
(74) 900 = 285 + 615
(75) 900 = 289 + 611
(76) 900 = 291 + 609
(77) 900 = 295 + 605
(78) 900 = 297 + 603
(79) 900 = 303 + 597
(80) 900 = 305 + 595

(100) 900 = 375 + 535
(101) 900 = 381 + 519
(102) 900 = 385 + 515
(103) 900 = 387 + 513
(104) 900 = 393 + 507
(105) 900 = 395 + 505
(106) 900 = 399 + 501

(107) 900 = 403 + 497
(108) 900 = 405 + 495
(109) 900 = 407 + 493
(110) 900 = 411 + 489
(111) 900 = 415 + 485
(112) 900 = 417 + 483
(113) 900 = 423 + 477
(114) 900 = 425 + 475
(115) 900 = 427 + 473
(116) 900 = 429 + 471
(117) 900 = 435 + 465
(118) 900 = 441 + 459
(119) 900 = 445 + 455
(120) 900 = 447 + 453

X = K + p / 2 – d / 2 = 120 + 153 / 2 – 297 / 2 = 48 **(Refer to Page 14)**

Assuming p does not exist after N = 500, then p_{500} = 94, d_{500} = 156 + 200 = 356.

In the context of: X = K + p / 2 - d / 2
If X = 0, then: K = d / 2 – p / 2 = 356 / 2 – 94 / 2 = 131
Let: K = K_1 + K_2, and K_1 represents the value when N = 400, p_{400} = 77
Then: K_1 = N / 2 – p = 400 / 2 – 77 = 123
But: K_2 = 14 **(see the last 14 pairs of K above)**

Verification of K_2:

(1) 900 = 403 + 497
(2) 900 = 405 + 495
(3) 900 = 407 + 493
(4) 900 = 411 + 489
(5) 900 = 415 + 485
(6) 900 = 417 + 483
(7) 900 = 403 + 477
(8) 900 = 425 + 475
(9) 900 = 427 + 473
(10) 900 = 429 + 471
(11) 900 = 435 + 465
(12) 900 = 441 + 459
(13) 900 = 445 + 455
(14) 900 = 447 + 453

K = K_1 + K_2 = 123 + 14 = 137
X = K + p / 2 - d / 2 = 137 + 94 / 2 - 356 / 2 = 6 **(Refer to Page 14)**

(1) 900 = 401 + 499
(2) 900 = 409 + 491
(3) 900 = 421 + 479
(4) 900 = 433 + 467
(5) 900 = 439 + 461
(6) 900 = 443 + 457

Conclusion: There are 6 pairs comply with the (GC) when we assume no prime numbers exist after N = 500, but actually there are 48 pairs.

[Situation 3]: Assuming that no prime numbers exist between N = 5000 and 6000, thus:

N = 6000 $p_{5000} = 668$ $d_{5000} = 2332$ $K_{5000} = 945$

$$
\begin{aligned}
X &= K + p / 2 - d / 2 \\
&= 945 + (668 / 2 - 2332 / 2) \\
&= 945 + (334 - 1166) \\
&= 945 - 832 \\
&= 113
\end{aligned}
$$

Verification of X:

(1) 6000 = 1013 + 4987
(2) 6000 = 1031 + 4969
(3) 6000 = 1033 + 4967
(4) 6000 = 1049 + 4951
(5) 6000 = 1063 + 4937
(6) 6000 = 1069 + 3931
(7) 6000 = 1097 + 4903
(8) 6000 = 1123 + 4877
(9) 6000 = 1129 + 4871
(10) 6000 = 1187 + 4813
(11) 6000 = 1201 + 4799
(12) 6000 = 1213 + 4787
(13) 6000 = 1217 + 4783
(14) 6000 = 1249 + 4751
(15) 6000 = 1277 + 4723
(16) 6000 = 1279 + 4721
(17) 6000 = 1297 + 4703
(18) 6000 = 1321 + 4679
(19) 6000 = 1327 + 4673
(20) 6000 = 1361 + 4639
(21) 6000 = 1409 + 4591
(22) 6000 = 1433 + 4567
(23) 6000 = 1439 + 4561
(24) 6000 = 1451 + 4549
(25) 6000 = 1453 + 4597
(26) 6000 = 1481 + 4519
(27) 6000 = 1483 + 4517
(28) 6000 = 1487 + 4513
(29) 6000 = 1493 + 4507
(30) 6000 = 1343 + 4457
(31) 6000 = 1549 + 4451
(32) 6000 = 1553 + 4447
(33) 6000 = 1559 + 4441
(34) 6000 = 1579 + 4421
(35) 6000 = 1609 + 4391
(36) 6000 = 1627 + 4373
(37) 6000 = 1637 + 4363
(38) 6000 = 1663 + 4337
(39) 6000 = 1741 + 4259
(40) 6000 = 1747 + 4253
(41) 6000 = 1759 + 4241
(42) 6000 = 1783 + 4217
(43) 6000 = 1789 + 4211
(44) 6000 = 1823 + 4177
(45) 6000 = 1847 + 4153
(46) 6000 = 1861 + 4139
(47) 6000 = 1867 + 4133
(48) 6000 = 1871 + 4129
(49) 6000 = 1873 + 4127
(50) 6000 = 1889 + 4111
(51) 6000 = 1901 + 4099
(52) 6000 = 1907 + 4093
(53) 6000 = 1949 + 4052
(54) 6000 = 1951 + 4049
(55) 6000 = 1973 + 4027
(56) 6000 = 1979 + 4021
(57) 6000 = 1987 + 4013
(58) 6000 = 1993 + 4007
(59) 6000 = 1997 + 4003
(60) 6000 = 1999 + 4001
(61) 6000 = 1201 + 3989
(62) 6000 = 2053 + 3947
(63) 6000 = 2069 + 3931
(64) 6000 = 2081 + 3919
(65) 6000 = 2083 + 3917
(66) 6000 = 2089 + 3911
(67) 6000 = 2111 + 3889
(68) 6000 = 2137 + 3863
(69) 6000 = 2153 + 3847
(70) 6000 = 2179 + 3821
(71) 6000 = 2203 + 3797
(72) 6000 = 2207 + 3793
(73) 6000 = 2221 + 3779
(74) 6000 = 2239 + 3761
(75) 6000 = 2267 + 3733
(76) 6000 = 2273 + 3727
(77) 6000 = 2281 + 3719
(78) 6000 = 2309 + 3691
(79) 6000 = 2341 + 3659
(80) 6000 = 2357 + 3643
(81) 6000 = 2377 + 3623
(82) 6000 = 2383 + 3617
(83) 6000 = 2393 + 3607
(84) 6000 = 2417 + 3583
(85) 6000 = 2441 + 3559
(86) 6000 = 2459 + 3541
(87) 6000 = 2467 + 3533
(88) 6000 = 2473 + 4591
(89) 6000 = 2531 + 3469
(90) 6000 = 2539 + 3461
(91) 6000 = 2543 + 3457
(93) 6000 = 2593 + 3407
(94) 6000 = 2609 + 3391
(95) 6000 = 2657 + 3343
(96) 6000 = 2671 + 3329
(97) 6000 = 2677 + 3323
(98) 6000 = 2687 + 3313
(99) 6000 = 2693 + 3307
(100) 6000 = 2699 + 3301
(101) 6000 = 2729 + 3271
(102) 6000 = 2741 + 3259

(103) 6000 = 2749 + 3251
(104) 6000 = 2791 + 3209
(105) 6000 = 2797 + 3203
(106) 6000 = 2819 + 3181
(107) 6000 = 2833 + 3167
(108) 6000 = 2837 + 3163
(109) 6000 = 2879 + 3121
(110) 6000 = 2917 + 3083
(111) 6000 = 2939 + 3061
(112) 6000 = 2963 + 3037
(113) 6000 = 2999 + 3001

So, although we are knowledgeable if there are any prime numbers in the future (after 5000), we are still able to find 113 pairs of primes --- such that the sum of each pair is equal to 6000 (N).

Hence, given even number $N \geq 6$, both p and d can be determined. Consequently, K and X can be resolved, and the boosting K value has ensured the increasing nature of X value. Based on the foundation of N = 116, K = 6, X = 6 for our discussion, notice that our discussion is no longer "conjecture" but "certainty".

6. THE WEAK CONJECTURE

"A Weak Goldbach's conjecture" is known as --- all the odd number ≥ 9 can be expressed as the sum of three odd primes.

Since 9 = 3 + N, and N can be represented by two odd primes; naturally 9 will be the sum of three primes, when N becomes larger, the odd numbers will be larger too. This is supported unambiguously by a graphic proof **(Appendix D)** in CHAPTER FOUR.

7. THE CONCLUSION

(1) Goldbach's conjecture --- the principal problem in number theory is "arithmetic" in nature; it can be understood through studying the "patterns of the whole numbers" and "elementary calculation techniques" --- via reasoning.
(2) The general formula $X = K + p / 2 - d / 2$ was discovered in the range of N = 40, but the logic led to its formulation is more complicated (available if needed)

External links are retrieved for the general information in number theory and Goldbach's conjecture only.

[1]: Courant R. and Robbins H. "What is mathematics?" 15th edition, 1973
[2]: G.H Hardy; E. M. Wright (2008) [1938]. An introduction to the theory of numbers (rev. by D. R Heath-Brown and J. H Silverman, 6th ed) Oxford University Press. ISBN 978-0-19-9211986-5 http://google.com/book?id = rey9wfSaJ9EC&dq
[3]: Title: EXIST/ Sheng, Shi Feng
Copyrighted with the Library of Congress
Registration Number & Date: TXU0003465073/1990 – 12 – 4.
Contents: Goldbach's conjecture, Fermat's Last Theorem, an infinite calculation chart, trisecting the angle, heptagon, doubling the cube, squaring the circle.
ISBN 978-988-19795-1-3 (unpublished)

EXHIBIT B AN INFINITE CALCULATION CHART

It is imperative to have a deep understanding of:
*(**Exhibit A)** --- which proved the Goldbach's principle (1+1) --- to mean all the even number > 2 can be expressed as the sum of two prime numbers.
* A graphic study of the Goldbach's conjecture **(Appendix D)** in CHAPTER FOUR

THEN refer to chart **(Appendix E-1)** in CHAPTER FOUR --- this blank chart is a crystallization of Goldbach's conjecture. The 10 squares expressed the Goldbach's principle 1 + 1 + 1 + 1 + ... moreover; they represent the arithmetical progression of positive integer 1, 10, 100, 1000, 10000, 100000 ... As we know primes > 2 exist in odd numbers only, and both even and odd numbers exist in nature numbers. Infinitude of natural numbers implied that Goldbach's conjecture is interminable, which is comprehensively connoted by this chart made of boundless straight lines. It consists of two squares; we called the 1st square as **Chart I** and the 2nd square as **Chart II.**

This chart is truly an important discovery in mathematics. (No special mathematical knowledge is needed). If you are a smart person, you may find it useful to enrich your imagination and resolve some complicated problems. This chart is an ocean of knowledge. Upon becoming an experience user, you will be stuck by the fact that:

(1) It outperforms any computer, both in speed and range, however large or small:

$3^{50000} = 1.155407015 \times (10)^{23856}$
$1.11 / 246835792 = 4.48 \times (10)^{-10}$

(2) It is a rigorous calculation chart that uses colors, lines and spaces to form a mathematical language that performs object calculations to give instant answers and unified all the pre-Newtonian mathematics

(3) Amazingly, it also represents as **a rigorous infinite topological field** that gives information to resolve those ancient Greek straightedges and compass problems --- "trisecting an angle" to resolve "polygons", "doubling the cube", and "squaring the circle" declared by others and Courant and Robbins as **"insoluble"** in [8]. They will be elaborated in an appropriate time.

Composition of the chart (Use Chart E-2)

1. There are 10 vertical and horizontal **(red lines)** that represent the most significant digit of number **(1.....9).** There are 9 vertical and horizontal **(blue lines)** that lie between **(red lines)** represent the second most significant digit of number **(1.....9)**
2. In between the **(blue lines)** you will find one or four **(green lines).** In case of one, it indicates 5 (prime number). In the case of four, they take on **(2, 4, 6, 8)**
3. Those **(slanted parallel lines) and (radical lines)** will be explained later
4. The scales on four sides of the chart acted as the ruler to read the results of different calculation.

****

Chart E-2 is exactly same as **Chart E-1** --- except Y-axis was extended and it provides more information to follow the reading rules and examples below.

****

Reading Rules

(1) Always read data from left to right, and bottom to top.
(2) The (red lines) in **chart I** represent 1 to 9, and those in **chart II** represent 10 to 90
(3) The **(right-left slanted parallel lines)** are for multiplication and division. You read the same way as you deal with vertical lines.
(4) The **(left-right slanted parallel lines)** are for fraction operation. It is read from left bottom to up right. The first line represents 1, these lines indicate **numerator** and the vertical lines represent **denominator.**

Find 3.46 from horizontal lines
Count red horizontal lines bottom –up until you reach the third one, then count four blue lines in the same direction, finally move up three green lines

Find 4.25 from vertical lines
Count red vertical lines from left to right until you reach the fourth one, then count two blue lines, continue to move right to the only green line

I. Multiplication and Division:

(1) $2 \times 2 = 4$ (Multiply is equivalent to add)

a. Count red horizontal lines bottom –up until you reach the 2nd one.
b. Count forward slanted lines until you reach the 2nd one (count left to right)
c. Read the vertical line that passes through the intersection point, it showed 4 on the bottom scaled ruler.

(2) $4 \times 4 = 16$

a. Count red horizontal lines bottom –up until you reach the 4th one.
b. Count forward slanted lines until you reach the 2nd one (count left to right)
c. Read the vertical line that passes through the intersection point, it showed 16 on the bottom scaled ruler.

(3) $2.4 \times 3.2 = 7.68$

a. Find 2.4 from vertical lines (count bottom –up)
b. Find 3.3 from forward slanted lines (count left – right)
c. Read the vertical line that pass through the intersection point. The result should be 7.68

(4) $1.4 \times 3.6 = 5.04$

a. Find 1.4 from horizontal lines (count bottom up)
b. Find 3.6 from forward slanted lines (left to right)
c. Read the vertical lines that passes through the intersection point, and showed 5.04

(5) $2.4 \div 1.4 = 1.71$ (Division is equivalent to subtract)

a. Find 2.4 from vertical lines (count left to right)
b. Find 1.4 from forward slanted lines
c. Read the vertical lines that passes through the intersection point, and showed 1.71

(6) $2.4 \div 3.2 = 0.75$

a. Find 2.4 from vertical lines (count left to right)
b. Find 3.2 from forward slanted lines
c. these two lines do not intersect, so locate 24 in (chart II)
d. Now, they intersection at a point, that should reads 0.75 (horizontally)

II. Fraction (within chart I only)

(1) $7 / 3 = 2.33$

a. Count red vertical lines (left to right), the third one represents **denominator** 3
b. Count back slanted lines to find 7, note that the origin is 1
c. The intersection point should be 2.33 on the vertical axis

(2) $3 / 7 = 0.428$

a. count 7 vertical lines from origin, move up until it intersect the 2nd backward slanted line in (chart II) which indicates 30
b. The intersection point reads 4.28 on vertical axis, shift left one decimal point, we get 0.428.

III. Fractional Power

(1) $4^{½} = 2$

a. Find the vertical line that represents 4
b. Locate the 2nd radical line that swing around the lower left corner clockwise
c. The intersection of these two lines is the resultant point.
d. Read the corresponding horizontal line, it should be 2

(2) $4.2^{½} = 2.05$

a. Find the vertical line that represents 4.2
b. Locate the 2nd radical line that swing around the lower left corner clockwise

c. The intersection of these two lines is the resultant point.
d. Read the corresponding horizontal line, it should be 2.05.

(3) $22^{1/10} = 1.36$

a. Find the vertical line that represents 22 (in chart II)
b. Locate the 10th radical line that swing around the lower left corner clockwise
c. The intersection of these two lines is the resultant point.
d. Read the corresponding horizontal line, it should be 1.36.

(4) $5.4^{0.1} = 1.18$
Note: we need to work in the right half of the chart (Chart II)

a. Find the vertical line from middle to right that corresponds to 5.4
b. Locate the 1st radical line that swings counter clockwise from the base line. This corresponds to value 0.1 (red line)
c. The intersection of these two lines is the resultant point.
d. Read bottom –up the horizontal line that passes through this point, it should be 1.18

(5) $5.4^{0.7} = 3.26$ Note: we need to work in the right half of the chart (Chart II)

a. Locate the vertical line from middle to right that corresponds to 5.4
b. Locate the 7th radical line that swings counter clockwise from the base line. This corresponds to value 0.7 (red line)
c. The intersection of these two lines is the resultant point.
d. Read bottom –up the horizontal line that passes through this point, it should be 3.26

IV. Positive power (integer power)

(1) $3.4^2 = 11.6$

a. Find the value 3.4 from horizontal line (count bottom up)
b. Locate the 2nd radical line that wings around the origin, which represents the power value 2
c. The intersection of these two lines is the resultant point
d. Read the vertical line that proceeds through this point, it should be 11.6

(2) $2^5 = 32$

a. Find the value 2 from horizontal line (count bottom up)
b. Locate the 5th radical line that wings around the origin, which represents the power value 5
c. The intersection of these two lines is the resultant point
d. Read the vertical line that proceeds through this point. It should be 32

(3) Practice to locate:

$2^2 = 2 \times 2 = 2 + 2 = 4$ **(2 is the only even prime number)**
$2^4 = 16$
$2^6 = 64$
$3^4 = 81$

V. Logarithm (lg 10 = 1)

(1) Log 135 = 2.13

a. The integral part of the outcome is equal to the value that subtract one from the length of the number, so it should be 3 – 1 = 2 in this case.
b. Look at (Chart II), let the bottom red line represents 100, and count bottom up to find 135 and read the scaled vertical ruler on the right most side. The value contributes to the decimal part is 0.13.
c. Assemble the integer and decimal, we get 2 + 0.13 = 2.13

(2) Log 4561 = 3.659

a. the integral part of the outcome is equal to the value that subtract one from the length of the number, so it should be 4 – 1 = 3 in this case.
b. Look at (Chart II), let the bottom red line represents 1000. Find 4561 and read the scaled vertical ruler on the right most side. The value contributes to the decimal part should be 0.659.
c. Assemble the integer and decimal, we get 3 + 0.659 = 3.659

VI. Trigonometric functions

(1) Sin 10^0 30' = 1.182

a. Count horizontal lines bottom –up from the left most scales ruler to find 100 30,
b. read from the same ruler we get 1.82
c. Since sin x < = 1, so sin 10^0 30'

(2) Sin 1^0 30' = 0.0261

a. Look for 1^0 30' from the top ruler of the chart. Read the result we get 2.61.
b. Tan 45 ^{0} = 1, so sin 1^0 30' = 0, 0261.

(3) Tan 2 ^{0} = 0.0349

Both bottom and top scaled rulers are for function than x. just like the left and right rulers. Please practice to locate the following in the ruler:

Tan 10^0 = 0.176, Tan 55^0 = 1.43, Tan 86^0 20' = 15.6,

VII. Complex calculation

Sin {[(log2468135792) $^{1/10}$] $^{0.45}$ / 246835792}' = 1.30 × (10) $^{-13}$

a. log 2468135792 = 9.392
b. 9.392 $^{1/10}$ = 1.25
c. .25 $^{0.45}$ = 1.11
d. 1.11 / 246835792 = 4.48 × (10) $^{-10}$
e. Sin (4.48 × 10 $^{-13}$) = 1.30 × (10) $^{-13}$ (1^0 = 60')

tan {[(log 2975318642)$^{1/7}$]$^{0.65}$ / 2975318642}' = 1.21 × (10) $^{-13}$

a. log 2975318642 = 9.474
b. 9.474 $^{1/7}$ = 1.38
c. 1.38 $^{0.65}$ = 1.23
d. 1.23 / 2975318642 = 4.14 × (10) $^{-10}$
e. tan (4.14 × 10 $^{-10}$) = 1.21 × (10) $^{-13}$ (1^0 = 60')

VIII. To perform certain tasks that your calculator fails to perform:

3 ^{500000} = 4.239845195 × (10) ^{238560}
3 ^{50000} = 1.155407015 × (10) ^{23856}

2 ^{10000} = 1.995062934 × (10) ^{3020}
2 ^{1000} = 1.071508607 × (10) ^{302}

33 ^{5000} = 3.712778154 × (10) ^{7592}
33 ^{500} = 1.80748879 × (10) ^{759}

333333333 ^{50000} = 8.654460490 × (10) ^{426143}
333333333 ^{5000} = 2.475848123 × (10) ^{42614}

VIIII. to dispute the analytic number theory --- via accurate calculation:

In 1742, Goldbach's conjectured that:

(a) Any even number N ≥ 6 can be expressed as the sum of two odd prime numbers; (b) any odd number ≥ 9 can be expressed as the sum of three odd primes. It is also known as (1 + 1) --- to mean all the even numbers > 2 can be expressed as the sum of two odd primes.

Nevertheless, this principal problem in pure mathematics was accepted as a problem of analytic number theory in the 20th century, and consequently, Vinogradov proved in 1937 that every sufficiently large odd number is the sum of three primes" [1]

In the 1960s, Professor J.R Chen (Beijing University) proved (1+2) --- to mean from some sufficiently large even number N on, all the even numbers are the sum of one prime and a product of two primes but (Chen did not tell you what N is) [2]. In the 1970(s), Chen and his teammates disclosed --- $P_x(1, 2) \geq 0.67 \times C_x / (\log)^2$ and showed:

$$62 = [7 + (5 \times 11)]$$
$$100 = [23 + (7 \times 11)]$$

According to [4] and [5], consequently, the 1937 Vinogadov's calculation ---

$$N \geq 3^{3^{15}} \approx e^{e^{16.573}} \approx 3.25 \times (10)^{6846168}$$

Was reduced by Chen and Wong in 1989 as:

$$e^{e^{11.503}} \approx 3.33 \times (10)^{43000}$$

Nevertheless, based on this infinite calculation chart, we obtained:

$$e^{e^{11.503}} = 3.333594466 \times (10)^{43000}$$

In pure mathematics, rigorous speaking, (i) close to ≠ equal to, however close, (ii) sufficiently large is still a finite number, however large.

Conclusion:

*Professor Chen (1933-1996) did not prove the Goldbach conjecture; nevertheless, the mathematical community accepted his (1 + 2) as the closest proof to the (1 + 1), and named **asteroid 7681** to honor his achievement in 1996 --- mainly because his approach was consistent with the analytic number theory --- the work of Dirichlet (1805-1859) and Yuri Linnik (1915-1972) introduced in [6] and [7].*

*Since the mathematical community accepted that (1+2) was the closest proof to (1+1) based on Dirichlet and Yuri Linnik --- we can only infer that the analytic number theory was **flawed.***

External links are retrieved from:

[1]: Vinogradov.I. M. "Representation of an odd number as the sum of three primes" comptes rendus (Doklady) de l'Academie des Sciences de ru R S S 15 169-172 1937

[2]: Chen J. R "on the representation of a Large even integer as the sum of a prime and a product of at most two primes. [Chinese] j. Kexue Tongbao 17 (1966), page 385-386

[3]: Chen J. R "on the representation of a Large even integer as the sum of a prime and a product of at most two primes. Sci. Sinidca 16, 15-76. 1773

[4]: http:// en.wikipedia.org /wiki/Jing Run,chen

[5]: Goldbach conjecture --- from Wolfram Math world
[6]: http:// en.wikipedia.org /wiki/ Dirichlet
[7]: http:// en.wikipedia.org /wiki/Linnik
[8]: Courant and Robbins. What is mathematics? An elementary Approach to ideas and methods, 15th Editon 1973
[9]: G.H Hardy; E. M. Wright (2008) [1938]. An introduction to the theory of numbers (rev. by D. R Heath-Brown and J. H Silverman, 6th ed) Oxford University Press. ISBN 978-0-19-9211986-5 http:// google.com/ book?id = rey9wfSaJ9EC&dq
[10]: Title: EXIST/ Sheng, Shi Feng
Copyrighted with the Library of Congress
Registration Number & Date: TXU0003465073/1990 – 12 – 4.
Contents: Goldbach's conjecture, an infinite calculation chart, Fermat's Last Theorem, trisecting the angle, heptagon, doubling the cube, squaring the circle.
ISBN 978-988-19795-1-3 (But unpublished)

EXHIBIT C A PROOF OF THE FERMAT'S LAST THEOREM (FLT)

ABSTRACT: After resolved the Goldbach's conjecture via an unprecedented polynomial formula that enlightened to construct an infinite calculation chart elaborated in **(Exhibit A), (Exhibit B);** it became clear that the Fermat's Last Theorem (FLT) is a non-geometry-based problem (arithmetic in nature) that can be resolved by a polynomial procedure, and it is well founded when Fermat claimed he had a truly remarkable proof (by descent).

(FLT) is the first important problem in number theory originated from the 3rd century "Arithmetica". In the 1736s, Leonhard Euler investigated this problem extensively without any success, nevertheless, his work marked the (rebirth) of (FLT) as the beginning of modern number theory, therefore, a revisit of the early history that led to this transformation is warranted.

1. THE PROBLEM

In about 1637, French mathematician Pierre de Fermat (1601- 1665) left a margin note in his copy of "Arithemtica" that this indeterminate Diophantine equation $A^N + B^N = C^N$ has no solution when N is an integer > 2. Furthermore, Fermat noted that he had a truly remarkable proof by (descent) but there was very little margin left to elaborate --- it is known as the famous (FLT) [1]

2. EARLY HISTORY

Mathematicians have dealt with questions of finding and describing the intersection of algebraic curves in early developments of algebraic geometry long before --- the Pythagorean theorems (400 B.C), the Element (350 BC) introduced the 1st proof of infinitude of primes by abstract reasoning, Euclidean geometry and addressed the tangent line and Archimedes (250 BC).

"Arithmetica" (260 AD) introduced Diophantine geometry; it was a collection of problems giving numerical solutions of determinate and indeterminate equations. Diophantine studied integers [....]. Diophantine also studied rational points on curves and algebraic varieties. In other words, he showed how to obtain infinitely many of rational points satisfying a system of equations via a procedure that can be made into an algebraic expression (basic algebraic geometry in pure algebraic forms). Unfortunately, most of texts were lost or unexplained.

Historically, some of the roots of algebraic geometry date back to the work of Hellenistic Greeks (450 BC). The Delain problem --- "doubling the cube" and other related problems such as; "trisecting the angle", "polygons", "squaring the circle" --- they are also known as straightedge and compass problems (or topological problems).

---

1 (FLT) is not an enigma --- it is about the solution of an equation with the form of $A^N + B^N = C^N$; where N = 2, it is well known as a formula of the Pythagorean right triangle $A^2 + B^2 = C^2$ ($3^2 + 4^2 = 5^2$), but when N = 3, then $3^3 + 4^3 = (27 + 64) = 91$, which is not a cube of any positive integer. It is also about $1^1 + 1^1 = 2^1$, but $1^2 + 1^2 = (\sqrt{2})^2$. Moreover, it is about $(2 + 2)^2 = 2^2 + (2 \times 2 \times 2) + 2^2 = 16 = (4)^2$.

In the 17th century --- Galileo (1564-1642) was the father of "observational astronomy" and "modern science", a polymath in the field of mathematics, physics, engineering and natural philosophy:

In mathematics, Galileo applied the standard passed down from ancient Greeks and Fibonacci (1170-1250); but superseded later by the algebraic methods of Descartes.

Rene' Descartes (1596-1630) discovered that --- assuming by restrict ourselves to the "straightedge and compass" in geometry, it is impossible to construct segments of every length. If we begin with a segment of length 1, say, we can only construct a segment of another length if it can be expressed using integers, addition, subtraction, multiplication, division and square roots (as the golden ratio can). Thus, one strategy to prove that a geometric problem is impossible (not constructible) --- is to show that the length of some segment in the final figure cannot be written in this way. But doing so rigorously required the nascent field of algebra [....]. Descartes introduced his analytic geometry primarily to study algebraic curves (reformulation of classic works on conic and cube). Using Descartes' approach, the geometric and logical arguments favored by the ancient Greeks for solving geometric problems could be replaced by doing algebra.

During the same period, Pierre de Fermat (1601- 1665) independently developed analytic geometry to study the properties of algebraic curves (those defined in Diophantine geometry), which is the manifestation of solutions of system of polynomial equations. Fermat also addressed Diophantine approximate equality to find maxima for functions and tangent lines to curves. However, most of Fermat's work was in private letters or margin notes --- his 1637 (FLT) was discovered after his death.

In physics --- Galileo's theoretical and experimental work on the motions of bodies, along with the work of Kepler (1571-1630) and Descartes was a precursor of the classical mechanics developed by Isaac Newton (1642-1726). Initially, Newton developed calculus to study physical problems; it is a collection of techniques for manipulating infinitesimal and capable to approximate a polynomial series, but infinitesimals do not satisfy the Archimedean property.

Moreover; Leibniz (1646-1716) independently developed calculus. In geometry, Leibniz defined tangent line as a line through a pair of infinitely close points --- as one of the most fundamental notions in differential geometry.

In the 18th century, physicist/mathematician Daniel Bernoulli (1700-1782) accepted the 17th century analytic geometry *only because* it supplied with concrete quantitative tools needed to analyze his physical fluid motion problem via infinitesimal calculus; and his colleague Leonhard Euler (1707-1783) contributed greatly; it was in the same period:

(a) Euler became interested in number theory after Christian Goldbach (1690-1764) introduced Fermat's unelaborated work (including the FLT) to him.
(b) adapted the infinitesimal calculus and his zeta function to investigate prime distribution and obtained his finite products via a round-about method:

$$\zeta(s) = \sum_{n=1}^{\infty} 1 / n^s = 1 + 1 / 2^s + 1 / 3^s + 1 / 4^s + \ldots = \prod_p (1 - p^{-s})^{-1}$$

Although Euler was unsuccessful in all fronts, but his investigation of Fermat's work marked the "rebirth" of the (FLT) as the beginning of modern number theory.

3. OBSERVATION AND REASONING

(a) Hermann Hankel (1839-1873) studied "Arithemtica" extensively and commented --- not the slightest trace of general, comprehensive method is Discernible; each problem calls for some special method which refuses to work even for the most related problems. For this reason it is difficult for modern scholars to solve the 101st problem even after having studied 100 of Diophantus's solutions [2]

(b) Although Fermat's last Theorem was noted is his copy of "Arithmetica " but not slightest evidence suggest that finding infinite rational points on the curves was possible. However, Fermat's proof must be a short one --- since he implied it could be a margin note if he had a little more space.

(c) When we speak of Diophantine equations today, we are speaking of polynomial equations, in which integer solutions must be found. Polynomial is a mathematical expression that involves N's and N^2s and N's raise to other powers

(d) An integer $N > 2$ is equivalent to $N \geq 3$. Moreover, Fermat did no constrain A and B in equation $A^N + B^N = C^N$.

4. TO PROVE EQUATION $A^N + B^N = C^N$ HAS NO INTEGER SOLUTIONS by (descent) WHEN N IS AN INTEGER > 2

[Study 1]: Let: A = B = 2; and descend N from 3 → 1 and 2:

Then: $A^N + B^N = C^N$

Became: $2^N + 2^N = 4^N$ (A, B and C have a common prime factor 2)

N = 1: $2^1 + 2^1 = 4^1$ (4 = 4)

N = 2: $2^2 + 2^2 \neq 4^2$ (8 ≠ 16)

Because: $(2 + 2)^2 = 2^2 + (2 \times 2 \times 2) + 2^2 = 16$

It is undisputed that $2^2 + 2^2 = 2^3$ (4 + 4 = 8), and a certain polynomial algorithm (of a particular form) can be easily constructed as:

$2^N + 2^N = 2^{N+1}$ Let: N = 3, 4, 5 ...

N = 3: $2^3 + 2^3 = 2^4$ (8 + 8 = 16)

to hold Fermat's statement true infinitely that this equation $A^N + B^N = C^N$ has no solution when integer $N > 2$ because:

(i) 4, 6, 8 ... have a common prime factor 2, and (prime 2) is the building block of all odd prime and composite numbers. For example:

1 + 2 = <u>3</u> 3 + 2 = <u>5</u> 5 + 2 = <u>7</u> 7 + 2 = 9 9 + 2 = <u>11</u> 11 + 2 = <u>13</u>
13 + 2 = 15 15 + 2 = <u>17</u> 17 + 2 = <u>19</u> 19 + 2 = 21 So on to the forth

(ii) Naturally, infinitude of odd and prime numbers ≥ 3 have a common (prime factor 2) built in, therefore, all we need is to prove $A^N + B^N \neq C^N$ when $N = 2$

[Study 2]: Nevertheless, the following Pythagorean square illustrated equation $A^N + B^N = C^N$ exists when $N = 2$, but it is only limited to as the equation of Pythagorean right triangle.

Let A = B:

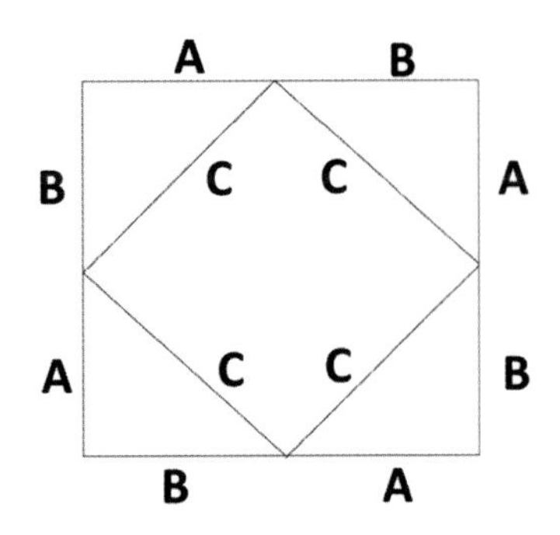

$(A + B)^2 = C^2 + 4[1/2\ (A \times B)]$
$A^2 + (2 \times A \times B) + B^2 = C^2 + (2 \times A \times B)$
$A^2 + B^2 = C^2 + (2 \times A \times B) - (2 \times A \times B)$
$A^2 + B^2 = C^2$

5. THE CONCLUSION

(FLT) was correct and it is well founded when Fermat claimed he had a truly remarkable proof by (descent) --- because this indeterminate equation $A^N + B^N = C^N$ could not be *solved* when integer $N > 2$, but the answer could be *verified* as a formula of Pythagorean right triangle when he descended integer N from $3 \rightarrow 2$ (a special case in mathematics).

External links are retrieved for general information of (FLT) only:

[1]: Title: EXIST/ Sheng, Shi Feng
Copyrighted with the Library of Congress
Registration Number & Date: TXU0003465073/1990 – 12 – 4.
Contents: Goldbach's conjecture, an infinite calculation chart, trisecting the angle, heptagon, doubling the cube, squaring the circle, Fermat's Last Theorem.
ISBN 978-988-19795-1-3 (But never published)

[2]: Hankel H; "*Geschichte der mathematic im alterumund mittelater,* Leipzig, 1874. (Translated to English by Ulrich Lirecht in Chinese mathematics in the 13th century, Dover publication, New York 1973

[3]: http://en.wikipedia.org/wiki/diophatus

EXHIBIT D TRISECTING THE ANGLE

Please use figure 1 (Appendix B) in chapter 4
An understanding of (chart E-1 in Exhibit B) is a must

This is one of the three "unsolvable" Greek problems. For centuries, it is believed that trisecting an angle with a straightedge and a pair of compasses alone is impossible, and people have taken that for granted. Now we have different opinions enlightened by **Goldbach's conjecture**. Remember, those 'insoluble' problems left by ancient Greeks are not foreordained facts, we have responsibility to crack the nut. In the following, let's try to put the matter into perspective and help you pare back the stance.

Subject: Two straight lines $\overline{OA}$, $\overline{OA'}$ intersect at point O, forming an angle $\angle AOA'$ (see Figure 1). Please trisect angle $\angle AOA'$ by ruler and compass only.

Procedures of Drawing

(1) Let A_1, A_2, A_3 be three points on line $\overline{OA}$ such that $\overline{OA_1} = 1/2\ \overline{OA_2} = 1/3\ \overline{OA_3}$. Using $\overline{OA_1}$, $\overline{OA_2}$, $\overline{OA_3}$ as radii, O as center of circles, draw three arcs $\widehat{A_1A_1'}$, $\widehat{A_2A_2'}$, $\widehat{A_3A_3'}$ through points A_1, A_2, A_3 and intersect line OA' at points A_1', A_2', A_3' respectively.

(2) Using points A_1 and A_2 as the centers of two circles, $\overline{OA_2}$ as radius, draw two arcs $\widehat{A_3D}$ and $\widehat{A_3'D}$, in which D is the intersection point. Link $\overline{OD}$, and extend to B. Line $\overline{OB}$ intersects arc $\widehat{A_3A_3'}$ (O as center, OA_3 as radius) at point F, and intersects arc $\widehat{A_2A_2'}$ (O as center, OA_2 as radius) at point E.

(3) Using point F as the center of a circle, $\widehat{EA_2}/2$ as radius ---------------- Draw a circle intersecting A_3D and $A_3'D$ at G and G respectively. Link $\overline{OG}$, $\overline{OG'}$ and extend to C and C' respectively. The intersection of $\overline{OC}$ and $\widehat{A_3A_3'}$ is H, and that of $\overline{OC'}$ and $\widehat{A_3A_3'}$ is H'. $\widehat{HH'} = 1/3\ \widehat{A_3A_3'}$, therefore $\widehat{HH'} = \widehat{HA_3} = \widehat{HA_3'}$, so $\angle AOC = \angle COC' = \angle C'OA'$.

(4) To make the drawing distinct, it is preferable to select an angle of appropriate size.

(5) This method can be adopted for the division of five equal angles or seven equal angles.

Proof:

Link $\overline{OG}$, and extend to C, The intersection of $\overline{OC}$ and $\overset{\frown}{A_3A_3'}$ is H, then using point G as the center of a circle, $\overline{GA_3}$ as radius, drawing a circle intersecting $\overset{\frown}{A_3A_3'}$ at H'. Link OH' and extend to C', G' is at $\overline{OH'}$

Additional Remarks:

Some mathematicians have theoretically proved the impossibility of any solutions to this problem. The methodology they employed is based on a trigonometric formula:

$$\text{Cos}\ \Theta = 4\ \text{Cos}^3(\Theta/3) - 3\text{Cos}(\Theta/3)\ \ldots\ldots\ldots\ldots(1)$$

Their conclusion stemmed from the following induction:

Let $z = \Theta/3$, $\Theta = 60°$, and $g = \text{Cos}\ \Theta = \text{Cos}\ 60° = 1/2$, equation (1) becomes:

$$4z^3 - 3z - g = 0\ \ldots\ldots\ldots\ldots\ldots(2)$$

Substitute g with 1/2, equation (2) becomes:

$$8\ z^3 - 6\ z = 1\ \ldots\ldots\ldots\ldots(3)$$

Let $v = 2z$, then equation (3) becomes:

$$v^3 - 3v = 1\ \ldots\ldots\ldots\ldots(4)$$

Assuming a rational number $v = r/s$ that satisfies this equation, in which both r and s are integers without a common factor that is greater than 1, we should have;

$r^3 - 3s^2r = s^3$. It can be inferred that $s^3 = r\ (r^2 - 3s^2)$ is divisible by r, or r and s must have a common factor unless $r = \pm1$. Similarly, s^2 is a factor of $r^3 = s^2\ (s+3r)$, or s and r must have a common factor unless $s^2 = \pm1$. Contrary to out assumption that r and s had no common factors, we can only adopt ±1 as the sole solution. Obviously neither +1 nor -1 satisfies equation (4), thus, equation (3) has no rational root. (See *What is Mathematics* by Courant and Robbins, 15th edition, 1973, pp. 137-138).

This approach is a mistake, because they entangled the fields of geometry and trigonometry, and ignored their individual exclusive characteristics. There are constraints to use equations, and the process should continue with a set of values, not just one. Let's see, when $\Theta = 60^\circ$, $\mathrm{Cos}\ \Theta = \mathrm{Cos}\ 60^\circ = 1/2$, and $\mathrm{Cos}(\Theta/3) = \mathrm{Cos}\ 20^\circ = 0.93969262...$, which implies that you can hardly expect to procure the result through solving above equations. Further, when $v = 2z = 2*(\Theta/3) = 1$, Θ should be 0°, you can not image an angle of 60° be divided into three 0° angles, nor can you rationalize that a 0° angle be divided into three 0° angles, although $\mathrm{Cos}\ 0^\circ = 4\ \mathrm{Cos}^3(0^\circ/3) - 3\ \mathrm{Cos}(0^\circ/3) = 1$.
In fact, there is always a solution to equation (1) provided that $\mathrm{Cos}(\Theta/3)$ is precise enough. Still let $\Theta = 60^\circ$

$$\mathrm{Cos}\ \Theta = 4\ \mathrm{Cos}^3(\Theta/3) - 3\ \mathrm{Cos}(\Theta/3)$$

$$\mathrm{Cos}\ 60^\circ = 4\ \mathrm{Cos}^3 20^\circ - 3\ \mathrm{Cos} 30^\circ$$

$$0.5 = 4*(0.93969262)^3 - 3*(0.93969262)$$

$$\underline{0.5 \approx 0.499999994}$$

Again, do not confuse these two areas when you study the math chart **(discussed in Exhibit B)**

As a practice, please refer to trisect this **60° (Appendix B)** in CHAPTER FOUR using a straightedge and a pair of compass only.

The principle is derived from the **Mathematical Chart** and the chart is the diagram of **Goldbach's Conjecture.**

EXHIBIT E REGULAR HEPTAGON

Please use diagram (Appendix B-1) in chapter 4
An understanding of the (trisecting the angle) is a must

This is the third of the three 'unsolvable' problems left by ancient Greeks. Now, based on our previous discussion, we can construct this regular heptagon by ruler and compass.

Procedures of Drawing

(1) Draw a circle and divide it into eight parts.
(2) Take one of the eight parts, then divide it into seven equal segments.
(3) Allot each of the seven segments to the rest seven parts, the result is seven equal sides inscribed in a circle. Please refer to *Trisect The Angle*. But here, I would like to offer a simpler method. Notice that 7/2 = 3.5, so if we add OA4 and find the half point of A_3A_4 at A3.5, then measure the arc $A_{3.5}A_{3.5}'$ with JA_1 or JA_1', we can get the seven points.
Please see diagram.
(4) Details of drawing:
 (i) Using OA_1, OA_2, OA_3, OA_4 as radii, draw arcs and intercept OA and OA' at A_1, A_2, A_3, A_4 and A_1', A_2', A_3', A_4' respectively. $OA_1 = OA_2/2 = OA_3/3 = OA_4/4$.
 (ii) Divide A_3A_4 and $A_3'A_4'$, we get points $A_{3.5}$ and $A_{3.5}'$. Using $OA_{3.5}$ or $OA_{3.5}'$ as radius, O as the center of the circle, draw arc $A_{3.5}A_{3.5}'$,
 (iii) Taking JA_1 or JA_1' as unit measure, divide arc A_2A_2, THEM SAME.
 (iv) You can check your drawing by measuring arcs A_1A_1 , A_2A_2', A_3A_3', A_4A_4' respectively.
 (v) Link relevant points, we get seven equal sides in the circle.

Explanation

It has been 'proved' that it is impossible to draw regular heptagon in a circle, but I believe that there is misunderstanding and confusion in the concept which leads to wrong conclusion. Please see *What Is Mathematics* by Courant and Robbins.

According to his theory, the seven vertices of the heptagon are represented by the roots of equation $x^7 - 1 = 0$. Each vertex's coordinate(x,y) corresponds to the real and imaginary part of the complex root z = x + yi. Obviously, one root of this equation is z = 1, the rest are roots of equation:

$$(z^7 - 1)/(z - 1) = z^6 + z^5 + z^4 + z^3 + z^2 + z + 1 = 0$$

Dividing this equation by z^3, we obtain the equation:

$$z^3 + 1/z^3 + z^2 + 1/z^2 + z + 1/z + 1 = 0$$

Change the form, we obtain

$$(z + 1/z)^3 - 3(z + 1/z) + (z + 1/z)^2 - 2 + (z + 1/z) + 1 = 0$$

Let $y = z + 1/z$, we find that

$$y^3 + y^2 - 2y - 1 = 0$$

We know that the seventh root(z) of unity can is given by

$$z = \cos\phi + i \sin\phi$$

Where angle $\phi = 360°/7$ subtends one edge of the regular heptagon, so $1/z = \cos \phi - i \sin \phi$ and $y = z + 1/z = 2\cos\phi$. If we can construct y, we can also construct $\cos \phi$, and conversely. Hence, if we can prove that y is not constructible, then we can not construct z and the heptagon neither. Thus, it is impossible to draw seven equal sides in the circle.

Assuming a rational root r/s, in which both r and s are integers without a common factor, then

$$r^3 + r^2s - 2rs^2 - s^3 = 0$$

It is seen from above that r^3 contains the factor s, and s^3 contains the factor r. Nevertheless, since s and r have no common factor, they must be ±1; therefore y can only be ±1. But neither of them satisfy above equation, hence, y is not solvable and consequently, the edge of a regular heptagon, is not constructible.

Now, let's give this approach a second thought.
We know y represents z+1/z and

$$z = \cos\phi + i \sin\phi$$
$$1/z = \cos\phi - i \sin\phi$$

Obviously, they are conjugate angles.
The basic concept is:

$$z = \rho (\cos \phi + i \sin \phi)$$

since $\rho = 1$, so $z = \cos\phi + i \sin\phi$

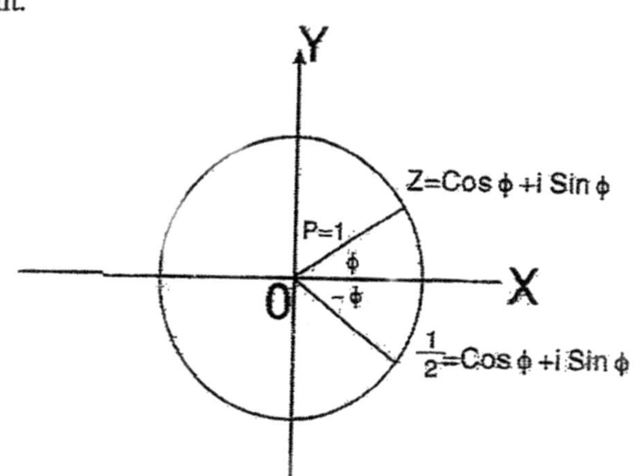

and
$y = z + 1/z$
$= (\cos\phi + i \sin\phi)+(\cos\phi - i \sin\phi)$
$= 2\cos\phi$

Remember our discussion is based on the equation:

$$y^3 + y^2 - 2y - 1 = 0$$

When $\rho = 1$, this is an unit circle, and z, 1/z are only two points on this circle.
How do we interpret the addition of these two points, i.e., $y = z + 1/z$. For a unit circle, the maximum and minimum values for x and y are +1 and -1. For $2\cos\phi$, if $\phi = 0$, $\cos\phi = 1$, and $y = 2\cos\phi = 2$, which exceeds the border of discussion. Thus, we are bound to stop if we expect $r/s = \pm1$ to satisfy equation $y^3 + y^2 - 2y -1 = 0$, because $y = 2\cos\phi$. If the result that a regular heptagon is not constructible is premised on a wrong assumption, how can we reach a correct conclusion? Even though we can draw y, there is still problem to draw "seven equal sides."

If we are able to draw $\phi = 360°/7$, the equation $y^3 + y^2 - 2y - 1 = 0$ can still be satisfied.
From $\phi = 360°/7 = 51.429°$ (regardless of preciseness)

$$y = 2\cos\phi = 2\cos 51.429° = 1.247$$
$$y^3 + y^2 - 2y - 1 = 0$$
$$1.247^3 + 1.247^2 - 2*1.247 - 1 = 0$$
$$3.494 - 3.494 = 0$$

Apparently, y satisfies the equation, but it can not be constructed because we do not understand the meaning of $z + 1/z$.

Up to now, we assert that the conclusion that a regular heptagon is not constructible is wrong. But this belief is so deeply ingrained in people's mind that it takes time to right the wrong. I suggest that you read my previous chapter which tells you how to divide an angle into three equal parts by rule and compass. Then the job to draw seven equal sides in a circle is as simple as measuring the body before cutting a cloth. For more reference, pleas see *Trisecting The Angle* and *Instant Calculation Chart.*

EXHIBIT F **DOUBLING THE CUBE**

Please use diagram **(Appendix C)** in chapter 4

An understanding of **(Chart E-1 in Exhibit B)** is a must

This is the second of the three 'unsolvable' Greek problems. Given a cube that has an edge of unit length, it is required to find the edge x of a cube with twice the volume of the unit cube. But this number x can not be constructed by ruler and compass. This conclusion is not correct, the following contents show you how to construct this edge.

Procedures of Drawing

(1) Draw two squares ABB'A' and BCC'B'. BB' coincide, AB, BC and A'B', B'C' are in alignment.

(2) Link AB', BC' and B'C.

(3) Link AC' and AD". AD" passes the intersection point of BC' and B'C. Note that there is a third imaginary square CDD'C' and AD" is part of AD'.

(4) Take AB as unit of measure, find log 2 of this unit and use it as radius to draw arcs using A and A' as the centers of circle to intersect AB and A'B' at two points respectively. Notice that these two points are symmetric and we marked "2" for the interception point on AB. Likewise, we obtain points "4", "8" and "16".

(5) Link every two points of the same mark value.

(6) The AD" is directly related to the doubling of the cube. Take "8" ($y_1 = 2$) as the base, we can easily find y_2 for which 16 is the doubling of the cube by linking two "16" points. This line intersects AD" at point P. The line segment from "16" on AB to P is the value of y_2. (See ΔMNP)

Proof

(1) log AB = log A + log B
If B = A, then log A*A = log A + log A
log A*A*A = log A + log A + log A

(2) Principle of design for this diagram:
X axis is 3log A.
Y axis is log A.
Therefore, if a triangle has sides parallel with X and Y axis, the relation can be expressed as:

$y/x = \log 10/ 3\log 10$ but $\log 10 = 1$
so, $y / x = 1 / 3$, $x = y^3$

(3) Consider $y_1 = 2$, $x_1 = 8$, when $x_2 = 16$, $y_2 = ?$
$y_2/y_1 = 16/8 = 2$ $y_2:y_1 = 2:1$

In fact, the geometric measure of y_2 in the diagram is 0.401, but the numerical measure is 2.52. So, $y_2^3 = 2.52^3 = 16$. The result complies with 16:8 = 2:1.

Discussion

Of course, this is another unresolved problem from ancient Greeks and it has also been wrongly destined 'insoluble' by Courant in his book *What is mathematics*. This is a mistake that should be put right.

Statement of the problem --- When x is the edge of a cube, it's volume is twice as much as that of a unit cube. Find this x using ruler and compass only. In algebraical expression: $x^3 = 2$
The two mathematicians mentioned in the book that there is only one root of real number for this equation, other roots are imaginary numbers. Since they know the fact that there is a real number solution, why did they propose an anti-fact conclusion?

In fact, the issue posed here is clear:

(i) A basic fact
The equation $x^3 -2 = 0$ has a root of real number, and this real number is the solution. It is another problem how we procure this number.

(ii) The methodology adopted by the two mathematicians has nothing to do with whether or not the equation $x^3 - 2 = 0$ has solution.

(iii) Their proof is a mistake:
According to their method, $2^{1/3}$ is an irrational number. Hence x can only lie in some extension field F_k, where k is a positive integer. So x can be written in the form $x = p + q\sqrt{w}$, p,q,w belong to some field F_{k-1}, but $\sqrt{w}$ does not. Similarly, $y = p - q\sqrt{w}$ is also a root of $x^3 - 2 = 0$
Since x belongs to the field F_k, so x^3 and $x^3 - 2$ are also in the this field F_k. Thus,

(2) $x^3 - 2 = a + b\sqrt{w}$

where a and b are in field F_{k-1}. After a simple calculation, we obtain:

$a = p^3 + 3pq^2w - 2$, $b = 3p^2q + q^3w$.

Let $y = p - q\sqrt{w}$, substitute q with -q in a and b, we get

(2') $y^3 - 2 = a - b\sqrt{w}$

Since we have assumed that x is a root of $x^3 - 2 = 0$, so

(3) $a + b\sqrt{w} = 0$

This implies that both a and b must be zero, otherwise from (3) we know $\sqrt{w} = -a/b$. $\sqrt{w}$ will become a number of F_{k-1} because a,b are in this field, thus b = 0 and from (3), a = 0.

From (2') we know $y = p - q\sqrt{w}$. As $y^3 - 2 = 0$ and $y \neq x$ i.e., $x - y \neq 0$ and $x - y = 2q\sqrt{w}$, if q = 0, then $x - y = 0$ and $x = p$ would lie in the field F_{k-1}. This is in contradiction with our assumption.

Consequently, the result shows that $x = p + q\sqrt{w}$ is a root of equation (1) and so is $y = p - q\sqrt{w}$. This outcome is conflicting with our assumption that there is only one real number as the cube root of 2, the other cube roots of 2 being imaginary numbers. Therefore, it is impossible to construct a doubling cube by ruler and compass.

This proof contains mistakes

(I) $x = p + q\sqrt{w}$ is not a real-number root of $x^3 - 2 = 0$

(II) $y = p - q\sqrt{w}$ is not a real-number root of $x^3 - 2 = 0$

(III) From mathematical point of view, $y^3 - 2 = 0$ can not be drawn from $y = p - q\sqrt{w}$.

(IV) When q = 0, x = p, y = p. They are the same number, not two different roots, let alone to say that the result causes contradiction.

(V) If $x = p + q\sqrt{w}$ and $y = p - q\sqrt{w}$ are roots of $x^3 - 2 = 0$, they must be conjugate imaginary numbers.

(VI) As is known, $x = p + q\sqrt{w}$ and $y = p - q\sqrt{w}$ are imaginary numbers expressed as a general solution for quadratic equation $Ax^2 + Bx + C = 0$. It is, therefore, useless to prove that they can not be the real-number roots for $x^3 - 2 = 0$ simultaneously.

In fact, the cubic equation $x^3 - 2 = 0$ can be resolved in a very simple way:

For (1) $x^3 - 2 = 0$

Change to $x^3 - c^3 = 0 \quad (c = 2^{1/3})$

Then we get $x^3 - c^3 = (x-c)(x^2 + cx + c^2) = 0$

So (i) $x - c = 0 \quad x = c$ i.e. $x = 2^{1/3}$

(ii) $x^2 + cx + c^2 = 0$

$$x = \frac{-c \pm \sqrt{c^2 - 4c^2}}{2} = -\frac{c}{2}(1 \mp \sqrt{3}i)$$

Thus,

$$\begin{cases} x = -\frac{c}{2}+\frac{c}{2}\sqrt{3}i \\ y = -\frac{c}{2}-\frac{c}{2}\sqrt{3}i \end{cases} \qquad \textit{Contrast with} \qquad \begin{cases} x = p+q\sqrt{w} \\ y = p-q\sqrt{w} \end{cases}$$

We get $p = -c/2$ $\quad q = c/2$

$\sqrt{w} = \sqrt{3}\, i$ $\quad c = 2^{1/3}$

The problem has turned out to be very simple and I believe that the above discussion has articulated those issues complicated by these two mathematicians. The key is to right the wrong conclusion of 'unsolvability'.

EXHIBIT G SQUARING THE CIRCLE

Please use diagram **(Appendix C-1)** in chapter 4

An understanding of **(Chart E-1 in Exhibit B)** is a must

This problem requires to construct a square which has an equal area of a circle with radius r. Since a circle's area is expressed as πr^2, the key to solving this problem is whether $\sqrt{\pi}$ is constructible. The following discussion provide a method of drawing this edge.

Procedures of Drawing

(1) Draw two squares ABB'C' and BCC'B'. BB' coincide, AB, BC and A'B',B'C' are in alignment.

(2) Link AB' and AC'. AC' is the diagonal line of the diagram.

(3) Take AB as unit of measure, find log 2, log 3 and log 8.5 of this unit. Then use log 2, log 3 as radii, A and A' as the centers of circle to draw arcs intercepting AB and A'B' at two sets of points. With these measures as bases, find "2", "4", "8" as well as "3", "6", "9" and "5". "16", "18", "17" are corresponding interception points.

(4) The two squares are identical in shape, but the second one is ten times the first one in scale. Thus, for "16", "17", "18" in the first square, their values are "1.6", "1.7", "1.8" respectively.

(5) Find log π, then draw a line vertical to AB and intercept AC' at point which makes value $\sqrt{\pi}$.

(6) According to D = 2 (R = 1), draw a circle. Use $\sqrt{\pi}$ to draw a square.

Proof

(1) log AB = log A + log B
If B = A, then log A*A = log A + log A

(2) Principle of design for this diagram:
X axis is 2log A.

Y axis is log A.
Therefore, if a triangle has sides parallel with X and Y axis, the relationship with original triangle can be expressed as:
$y/x = \log 10/ 2\log 10$ but $\log 10 = 1$
so, $y/x = 1/2$ $x = y^2$

(3) Let $x = \pi$, find y.
Evidently, $y^2 = \pi$, so $y = \sqrt{\pi}$ $(\log \sqrt{\pi} + \log \sqrt{\pi} = 2\log \sqrt{\pi})$
The circle area is:

$S = \pi R^2$ $R = 1$ $S = \pi$
$S = \pi D^2/4$ $D = 2$ $S = \pi$

The square area is:

$S^1 = x^2 (y^2)$ Now, $S = y^2 = \pi$

In the diagram, $x = \pi$, $y^2 = \pi$ So their areas are equal.

EXHIBIT 1 A PROOF OF THE P = NP in (P vs. NP)

ABSTRACT: P = NP is the principle part of the P vs. NP (a major unsolved problem in computer science) introduced by Stephen Cook in 1971. (a) Computer was initially developed to do arithmetic, it grew out of a theoretical investigation into the mathematical concept of "computability", (b) it is worth to mention that author Stephen Cook took a course in computer science during his freshman year (1951) and was hooked after wrote a program together with his friend to test the Goldbach's conjecture (GC) ---that he changed his major from Double E to Mathematics, (c) the (GC) is the principle unsolved problem in number theory (or arithmetic) and P = NP is the principle unsolved problem in computer science.

The objective here is to claim --- uncovering the pattern of the odd primes to prove the Goldbach's conjecture in **(Exhibit A)** should constituted as a proof of the **"P = NP"** --- because Goldbach's principle (1+1) inspired to construct an infinite calculation chart elaborated in **(Exhibit B)** that led to resolve most of the problems outlined by Stephen Cook, if P = NP.

1. THE PROBLEM

Informally speaking, Stephen Cook implied that if P = NP is proven; then it may lead:

(a) To resolve certain calculation tasks that a computer fail to perform.
(b) To stunning consequence in cryptography.
(c) To resolve a traveling salesman's problem introduced by Karl Menger in the 1930s.
(d) To predict protein structure.
(e) To transform mathematics by allowing a computer to find a formal proof of any theorem which has a proof of a reasonable length, since formal proofs can be recognized in polynomial time. Example problems may include all the CMI problems

2. P vs. NP

Official details of "The complexity of theorem-proving procedures" and "P vs. NP problem" are elaborated in [1] and [2]. Simply put, P vs. NP is a major unsolved problem in computer science introduced by Stephen Cook in his 1971 paper "The complexity of theorem proving procedures". The standard model of his computability theory is the 1936 Turing machine. It asks if every problem whose solution can be efficiently checked by a computer can also be solved by a computer quickly. Stephen Cook's theoretical notion of "quick" is an algorithm that runs in polynomial time "P". In his computability theory, Stephen Cook introduced three different classes of problem as; the "P", the "NP" and the "NP-completeness":

(a) The class of "P" consists of all those decision problems that can be solved on a deterministic computer in an amount of time is polynomial in the size of the input.
(b) The class of "NP" consists of all those decision problems whose positive solutions can be verified in polynomial time given the right information, or equivalently, whose solution can be found in polynomial on a non-deterministic computer.
(c) The class of "NP-Complete" consists of all the hardest problems in "NP" that need exponential time to resolve that is more likely can not be solved. The concept of NP-Complete provided his theory with a powerful tool to analyze computational tasks.

(d) Other important resources addressed in his computability theory are time (how many steps it takes to solve the problem) and space (how much memory it takes to resolve a problem). Computer is deterministic and sequential, meaning they can only perform actions one step after another.

3. OBSERVATION

"P" --- which stands for deterministic polynomial time. Polynomial is a mathematical expression that involves N's and N^2 s and N's raised to other powers. Roughly, "P" is a set of easy problems that can be solved on a deterministic computer in an amount of time is polynomial in the size of the input.

(1) To uncover the pattern of odd primes **(Appendix A-1)** in CHAPTER FOUR via a polynomial time procedure should constituted as a fundamental proof of the P = NP. Moreover, proving the Goldbach's conjecture via a polynomial time procedure elaborated in **(Exhibit A)** is characteristics of the "P = NP":

(Page 10 and 11 in Exhibit A) illustrated a polynomial time procedure was discovered to hold the (GC) true --- via a technique of ascending and descending order of odd numbers **(vertically)** in the range of any N between 10 to 42; the same technique can be adapted to hold the (GC) true in a reasonable range say N = 100, but solution time is proportional to the size of N involved.

Nonetheless, a certain computer program today instructs a moderate fast computer that performs 1 million basic arithmetical steps per / sec to give answers very quickly.

"NP"--- which stands for non-deterministic polynomial time; it is the set of problems that there is no known way to find an answer quickly, but if one is provided with the information to show what the answer is, then it may be possible to verify the answer more quickly. This class of questions for which an answer can be verified in polynomial time is called "NP".

(Exhibit A) illustrated that the (GC) automatically became a "NP" problem because there is no known way to find the answers quickly via the same technique say in the range of N = 130, 140, 150, 200 --- but if the information is provided to show what the answers are, then it will be possible to verify the answer more quickly, the solution time is proportional to the N involved.

Nevertheless, the same moderate fast computer performs 1 million basic arithmetical steps per / sec --- give the correct answers quickly. Naturally, more time and space will be needed as N increases and solution time is proportional to the size of N involved.

"NP-Complete" --- which stands for the hardest problem in "NP" that is more likely can not be resolved. They are hard not because they are difficult, but they will take exponential time that made impossible to solve. Indeed, the same technique of ascending and descending order of odd numbers say in the range of N = 1000, 2000, 5000 will take exponential time.

The 1742 Goldbach's conjecture certainly qualified as a "NP-Complete" problem that is more likely cannot be solved; it will take exponential time and space to obtain every pairs to hold the (GC) true infinitely --- since in mathematics, "infinity" is a greater value than any specified answer, however large.

The same moderate fast computer that performs 1 million basic arithmetical steps per / sec holds the (GC) true for a very, very large N --- the solution time is proportional to the size of N involved, naturally, it will take a computer exponential time and space --- but this is not a mathematical proof.

(2) According to "The Millennium Problems" [3]:

* "NP-Complete" was named by Stephen Cook after a curious property in a particular problem **(no name was mentioned).**

* "NP" problem is said to be a "NP-Complete" if the discovery of a polynomial time procedure "P" to solve it would imply that every "NP" problem could be solved by a polynomial time procedure.

* To identify a "NP-complete problem" and find the only "P" an algorithm runs in a polynomial time to solve the problem, then it is an automatic proof the "P = NP".

Please notice the general formula $X = K + p / 2 - d / 2$ in **(page 11 of Exhibit A)** --- it was the only possible "P" that runs in polynomial time --- to hold the Goldbach's statement true infinitely via a unique technique and step-by-step reasoning; moreover, the Goldbach's principle (1 + 1) enlightened to map out a **(Chart E-1 in Exhibit B)** to answer questions outlined below by Stephen Cook, if P = NP:

****

(a) With respect to resolve certain calculation tasks that a computer failed to perform;

Page 122 of "The millennium problems"[3] --- gave the comparisons of the times it would take a computer that perform (1 million basic arithmetical steps per / sec) to run processes having various time-complexity functions:

Time Complexity Function	Size of: N 10	20	30	40	50
N^2	0.0001 sec	0.0004 sec	0.0009 sec	0.0016 sec	0.0036 sec
N^3	0.001 sec	0.008 sec	0.027 sec	0.064 sec	0.125 sec
2^N	0.001 sec	1.0 sec	17.19 min	12.7 days	35.7 yrs
3^N	0.059	sec 58 min	6.5 yrs	3.85 centuries	200 million centuries

For the first two rows the processes run in polynomial time. But the last two rows show processes that run in exponential time. This table shows the enormous gulf that separates polynomial time processes from exponential processes [3]

Today's calculator (*TI-30Xa) and (Caliber)* --- obtained the solution of 3^{40} to 3^{200}, but **(Chart E-1 in Exhibit B)** provide accurate answers to show their approximations:

	Tl-30Xa (1990?)	Caliber (2000?)	(Chart E-1)
3^{40}	$1.215766546 \times 10^{19}$	$1.215766546 \times 10^{19}$	$1.2157665\underline{\mathbf{45}} \times 10^{19}$
3^{50}	$7.178979877 \times 10^{23}$	$7.178979877 \times 10^{23}$	$7.1789798\underline{\mathbf{61}} \times 10^{23}$
3^{200}	$2.656139889 \times 10^{95}$	$2.656139889 \times 10^{95}$	$2.6561398\underline{77} \times 10^{95}$
3^{300}	Error	Error	$1.368914772 \times 10^{143}$

(Tl-30X) and (Caliber) were flawed because their answer was obtained by calculating $3^{40} = (3^{30})^{1.333333333} = 1.215766546 \times 10^{19}$; but **(Chart E-1)** was mapped out based on polynomial process 1 + 1 + 1 + ... so the tendency is that their approximation will become bigger as N continues to increase.

Moreover, **(Chart E-1)** instructs an old Casio Fx-3600P calculator (built based on mathematical progression 10, 100, 1000 ...) to perform accurate calculations that outperform any computer both in speed and range, however large or small. For example:

$3^{500000} = 4.239845195 \times (10)^{238560}$
$3^{50000} = 1.155407015 \times (10)^{23856}$
$3^{5000} = 4.038996715 \times (10)^{2385}$

$333333333^{63000.51} = 4.242218172^{32941}$
$333333333^{63000.5} = 1.454652642^{3294}$

$99999999^{63000.51} = 1,201489556 \times (10)^{504004}$
$99999999^{63000.5} = 2.558420934 \times (10)^{50400}$

$\tan [(\log 2975318642)^{1/7}]^{0.65} / 2975318642\}' = 1.21 \times (10)^{-13}$
$1.11 / 246835792 = 4.48 \times (10)^{-10}$

(b) With respect to the cryptography:

Encryption scheme relies on certain problems being difficult. Researchers say that a proof of the P = NP may lead to stunning consequence in cryptography.

According to [1] --- a constructive and efficient solution to the NP-complete problem 3-SAT would break many existing cryptosystems such as Public-key cryptography, used for economic transactions over the internet, and Triple DES, used for transactions between banks.

These would need to be modified or replaced. The decryption key used in RSA method consists of two large prime numbers (each having 100 digits or more) chose by computer based on there is no known quick method of factoring large numbers.

** We are not knowledgeable in cryptography, but* ***(Chart E-1 in Exhibit B)*** *instructs a Casio fx-3600 calculator to factor any numbers quickly, however large. For example:*

23√ 916, 748, 679 .... a 201 digit numbers = 546,372,891 was resolved by mind calculator Shankuntala Devi (1929- 2013) in 50 seconds, and confirmed by a prepared Univac 1180 in 60 seconds in [4] --- but it will only take us about 30 seconds or less

(c) Computer scientists say to resolve a certain traveling salesman's problem is a proof of the P = NP. Please refer to the attached [**(Appendix 1A)**] for details

(d) Researchers say protein structure is a "NP-complete" problem, and if P = NP is proven; then it may lead to protein structure predation (psp). Proteins are essential to life, according to the latest published articles:

Structural Biology in "Science" page 478 / Vol 373/ Issue 6554 / 7-30-2021 --- announced that two groups of researchers unveiled the culmination of years of work by computer scientists, biologists and physicists: advanced modeling programs that can predict the precise 3D atomic structures of proteins. One team used (AI) programs to solve the structure of 350,000 from humans and 20 model organisms, such as E-Coli, yeast and fruit flies, all mainstays of biological research [....]

Computational Biology in "Nature" page 487-488 / Vol 596 / 8-26-2021 announced, and published two accepted research papers [5] and [6] in page 583-596.

** We are not in the field of biology but nature is relentless and unchangeable; if the computational method in [5] and [6] were on the right track; then this proof of the P = NP should be a great interest to the researchers in these two articles.*

(e) Stephen Cook implied that if P = NP is proven, then it will transform mathematics by allowing a computer to find a formal proof of any theorem which has a proof of a reasonable length, since formal proofs can be recognized in polynomial time. Example problems may include all the CMI problems. ***Indeed, please see the next seven (EXHIBITS)***

(3) In retrospect, the work of Alonzo Church [7] and Church -Turing Thesis (19J6) were the precursor of [1].

External links are retrieved for information related to P vs. NP only:

[1]: "The complexity of theorem-proving procedures" by Stephen Cook
[2]: The CMI official "P vs. NP problem"
[3]: P vs. NP in "The Millennium Problems" p.105- 129, p 110, p112, 215
[4]: http://en.wikipedia.org/wiki/shakuntala-Devi
[5]: https://doi.org/10.1038/s41586-021-03819-2 (Accurate psp with AlphaFord).
[6]: https://doi.org/10.1038/s41586-021-03828-1 (Accurate psp for the humanproteome)
[7]: "an unsolvable problem of elementary number theory" Church 1936
[8]: "Polynomial-time algorithm for prime factorization & discrete logarithms on quantum computer by Shor.P (1997)

[(Appendix 1A)] **A PROOF OF THE P = NP via the traveling salesman's problem**

ABSTRACT

Computer scientist said --- the traveling salesman problem introduced by mathematician Karl Menger in the 1930s is "a NP-complete problem"; so if this problem can be resolved by a polynomial time procedure "P", then it would follow that "P = NP". After proving the Goldbach's conjecture inspired to map out an infinite calculation chart (also represents as an infinite topological field) elaborated in **(Exhibit A)** and **(Chart E-1 in Exhibit B),** it became clear that --- this traveling salesman's problem can be resolved from a topological point view, and the same polynomial time procedure can be adapted to visit any number of cities.

1. P vs. NP

P vs. NP is a major unsolved problem in computer science introduced by Dr. Stephen Cook in 1971. The standard model in his computability theory is the 1936 Turing machine. Informally speaking, "P" is a set of relatively easy problem, and "NP" is a set of very hard problem. So "P = NP" would mean that a very hard problem apparently has a relatively easy answer. But the details are more complicated. Cook introduced three different classes of problems:

* The class "P" consists of all those decision problems that can be solved on a deterministic computer in an amount of time is polynomial in the size of the input.

* The class "NP" problem consists of all those decision problems whose positive solutions can be verified in polynomial time given the right information, or equivalently, whose solution can be found in polynomial time on a non-deterministic computer.

* The class of "NP-complete problem" consists of all the problems in "NP" that is more likely can not be solved because it will take exponential time to solve.

* Other important resources addressed in Cook's computability theory are time (how many steps it takes to solve the problem) and space (how much memory it takes to resolve a problem).

2. PRELUDE TO PROBLEM

According to [3] --- A salesman plans to find the shortest route to visit three different cities, namely: "Old-town", "Midtown" and "Newtown" from Springfield (hometown) with given distance between each pairs of the cities below:

	Springfield	Old-town	Midtown	Newtown
Springfield	0	54	17	79
Old-town	34	0	49	104
Midtown	17	49	0	91
Newtown	79	109	91	0

The distance from Newtown to Old-town is 5 miles further than the distance of from Old-town to Newtown because of a system of one-way street in Newtown. For all the other pairs of cities is same in both directions.

Mathematically, the shortest distance between two points is a straight line; so, find the shortest distance from (S) to the (1st city), then the shortest distance from (1st city) to the (2nd city), then the shortest distance from (2nd city) to the (3rd city) and return home can be easily add up as: S – M – O – N – S = 17 + 49 + 104 + 79 = 249. But computers are deterministic and sequential, meaning they can only perform actions one step after another. Therefore, the computer provides 3! = 3 × 2 × 1 = 6 options for the salesman listed below, and shortest route is 17 + 49 + 104 + 79 = 249

Route	Total mileage
S - O – M – N - S	54 + 49 + 91 + 79 = 273
S – O – N – M - S	54 + 104 + 91 + 17 = 266
S – M – N – O - S	17 + 91 + 109 + 54 = 271
S – M – O – N – S	17 + 49 + 104 + 79 = 249
S – N – O – M - S	79 + 109 + 49 + 17 = 254
S – N – M – O - S	79 + 91 + 49 + 54 = 273

3. THE PROBLEM

This salesman's problem comes when he plans to visit 10 cities (with given distance between each pair of cities); it automatically became a "NP-complete problem" because --- the computer provides 10! = 10 × 9 × 8 × 7 × 6 × 5 ×4 × 3 × 2 × 1 = 3.628.800 options that will take a computer exponential time (steps) and space (memory) that made impossible to find the shortest distance quickly.

4. OBSERVATION AND REASONING

Recap: (a) This salesman's problem is a NP-Complete, (b) Cook's theoretical notion of "quick" is an algorithm "p" that runs in polynomial time, (c) computer scientists today believed that there is no algorithm "p" can be found to resolve this problem, (d) if a "p" runs in polynomial time can be found to resolve this problem quickly; then it is automatic proof of P = NP, (e) polynomial is a mathematical expression that involves N's and N^2 s and N's raise to other powers.

(1) Researchers say topological space show up naturally in almost every branch of mathematics and this has made topology one of the great unifying ideas of mathematics.

*Indeed, please revisit (**Chart E-1 in Exhibit B**) -- it is a rigorous infinite calculation chart that uses colors, lines and spaces to form a mathematical language that performs object calculations and unified all the pre-Newtonian mathematics*

(2) Researchers say topology appears everywhere. Indeed, the subway map in N.Y.C is a form of non-rigorous topology that --- gives the relative position, closeness and connectedness between the stations.

Topologically, we can study this salesman's problem from a non-rigorous point of view, please refer to the 10 squares in ***(Chart E-1):***

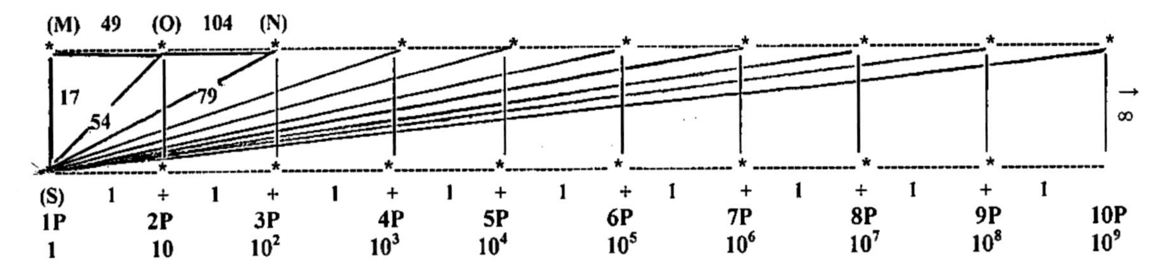

The 1st square represents visiting two cities, and the computer provides $2! = 2 \times 1 = 2$ options.

17 miles represents the shortest distance between (S) and (M).
49 miles represents the shortest distance between (M) and (O).
Topologically, shortest distance for the traveling salesman to visit two cities (M), (O) and return back to (S) is $17 + 49 + 54 = 120$ miles

It is a "P" problem that runs in polynomial time; it can be solved on a deterministic computer in an amount of time in the size of the input.

The 2nd square represents visiting three cities, and computer provides $3! = 3 \times 2 \times 1 = 6$ options:

104 miles represents the shortest distance between (O) and (N).
79 miles represents the shortest route returning from (N) to (S)
Topologically, shortest distance for the salesman to visit three cities and return home is $17 + 49 + 104 + 79 = 249$ miles.

It is a "NP" problem that can be verified in polynomial time given the right information, or equivalently, whose solution can be found in polynomial time on a non-deterministic, computer.

The 9th square represents visiting ten cities; and notices the exponential processes --- by just adding a few more cities:

$6! = 6 \times 5 \times 4 \times 3 \times 2 \times 1 = 720$
$7! = 7 \times 6 \times 5 \times 4 \times 3 \times 2 \times 1 = 5040$
$8! = 8 \times 7 \times 6 \times 5 \times 4 \times 3 \times 2 \times 1 = 40320$
$9! = 9 \times 8 \times 7 \times 6 \times 5 \times 4 \times 3 \times 2 \times 1 = 362880$
$10! = 10 \times 9 \times 8 \times 7 \times 6 \times 5 \times 4 \times 3 \times 2 \times 1 = 3.628.800$

To visit 10 cities automatically became a "NP-complete" problem --- because it will take a computer exponential time (steps) and space (memory) to print out 3.628.800 options quickly, but also will take this salesman exponential time to choose the shortest route that made his trip impossible.

But topologically, the shortest route to visit 10 cities can be easily resolved --- as long as the shortest distance between cities were given; moreover, the same polynomial time procedure can be adapted to visit N(s) cities --- since there are as many squares as numbers.

EXHIBIT 1-A A PROOF OF THE "P ≠ NP" IN P vs. NP

ABSTRACT

After proclaimed the proof of the P = NP in **(Exhibit 1)**, it became obvious that "A proof of the 1637 Fermat's Last Theorem (FLT) elaborated in **(Exhibit C)** should constituted as a proof of the P ≠ NP"

1. THE PROBLEM

The P ≠ NP in P vs. NP introduced by Stephen Cook in his 1971 paper "The complexity of theorem proving procedures" is still elusive today.

2. OBSERVATION

P vs. NP is a major unsolved problem in computer science introduced by Stephen Cook in his 1971 paper "The complexity of theorem proving procedures". The standard model of his computability theory is the 1936 Turing machine. It asks if every problem whose solution can be efficiently checked by a computer can also be solved by a computer quickly. Informally speaking, "P" is a set of relatively easy problems, and "NP" is a set of what seem to be very, very hard problems.

The official descriptions of "The complexity of theorem-proving procedures" and "P vs. NP problem" were elaborated in [1] and [2]. According to Stephen Cook:

(a) P ≠ NP would mean that some NP problems are harder to compute than to verify, they could not be solved in polynomial time, but the answer could be verified in polynomial.

(b) A proof show that P ≠ NP would lack the practical computational benefits of a proof of that P = NP, but would nevertheless represent a very significant advance in his computational complex theory and provide guidance for future research. It would allow one to show in a formal way that many common problems cannot be solved efficiently.

After CMI listed the 1971 P vs. NP as their millennium prize problem in 2000, many computer scientists believed that P ≠ NP. A key reason for their belief was that after decades of studying 3000 + known NP-complete problems, no one has been able to find a polynomial –time algorithm [...]

In 2002, Moshe Y. Vardi of Rice University stated that "the main argument in favor of P ≠ NP is the lack of fundamental progress in the area of exhaustive search. This is, in my opinion, a very weak argument. The space of algorithms is very large and we are only at the beginning of its exploration [...]

The resolution of Fermat's Last Theorem also shows that very simple questions may be settled only by very deep theories. Researchers spend their careers trying to prove theorems --- for instance, (FLT) took centuries to prove. A method that is guaranteed to find proofs to theorems, should one exist of a "reasonable size", would essentially end this struggle.

Moshe Vardi was referring to the 100+ pages proof of the (FLT) by Andrew Wiles in 1994, the general idea of Wiles was to claim a certain elliptic curve could not exist based on the 19th century existing knowledge of elliptic curves and the modern theory, theorem and conjecture of the rational points on elliptic curves by the 20th century abstractions of --- Poincare (1901), Mordell (1922), Mordell-Weil (1928), Hasse principle and Hasse-Weil (1940s) together with the work of others in the following chronological orders:

(a) Taniyama made a conjecture about elliptic curves in (1954)
(b) Frey postulates a connection between Taniyama conjecture and (FLT) in (1983)
(c) Kenneth Ribet proved the Frey connection in (1987).
(d) Wiles proved a special case of elliptic curve conjectured by Taniyama in (1993)

Nevertheless, Andrew Wiles' lengthy proof was unintelligible to the experts in the field, including Nick Katz [5], and [6] --- moreover, no one agrees that his proof was what Fermat had in mind since those deep theories did not exist during the Fermat's era; nevertheless, the mathematical community named asteroid **9999** to honor Wiles --- because his proof of the (FLT) was consistent with the existing knowledge.

3. THE CONCLUSION

$P \neq NP$ would mean that some NP problems are harder to compute than to verify, they could not be solved in polynomial time, but the answer could be verified in polynomial time. Hindsight, the short proof of the 1637 "Fermat's Last Theorem" elaborated in **(Exhibit C)** should constituted as a proof of the $P \neq NP$ --- because it clearly showed that (FLT) could not be solved in polynomial time, but the answer could be verified in polynomial time --- a special case in mathematics.

External links are retrieved for general information about (FLT) and P vs. NP only:

[1]: "The complexity of theorem-proving procedures" by Stephen Cook
[2]: The CMI official "P vs. NP problem"
[3]: http://en.wikipedia.org/wiki/P_versus_NP problem
[4]: http://www.claymath.org/millenumn/Pvs.NP / official problem
[5]: NYT VOL. CXLII.... No. 49372 copyright © 1993 The New York Times
[6]: "The world's most famous math problems" by Marilyn Vos Savant
[7]: Title: EXIST/ Sheng, Shi Feng
Copyrighted with the Library of Congress
Registration Number & Date: TXU0003465073/1990 – 12 – 4.
Contents:
Goldbach's conjecture, an infinite calculation chart, Fermat's Last Theorem, Exhibit D: trisecting the angle, Heptagon, doubling the cube, squaring the circle

EXHIBIT 2 A DISPROOF OF THE RIEMANN HYPOTHESIS (RH)

ABSTRACT: Historically, there is no known formula that yields all primes and no composites because primes are not polynomial. The (RH) refers to the pattern of the primes, but it (was) is unintelligible to the experts in the field. However, this conjecture re-appeared as a 2000 CMI millennium problem --- because the 20th century mathematician, physicist and computer scientist theorizing that if (RH) turns out to be correct, it will not only lead to understand of our infinite counting numbers system but also have implications in modern mathematics, physics, cryptography and beyond. Nevertheless, we disputed the validity of (RH) based on: (a) proving the Goldbach's conjecture elaborated in **(Exhibit A)** was credible, (b) the notion that Riemann Hypothesis has connection with physics, cryptography and others were speculations only.

1. THE PROBLEM

(RH) refers to the pattern of the primes, Riemann (1826-1866) was seeking to explain where every prime to infinity will occur by means of his analytic formula $\zeta(s) = 0$. Riemann approached by extending Euler's zeta function

$$\zeta\,(s) = \sum_{N=1}^{\infty} 1 / n^s$$

to the entire complex plane via analytic continuation with his original goal being to prove the validity of Gauss density conjecture. Riemann's $\zeta(s)$ extends to **C** as a complex function with only a simple pole at s = 1, with residue 1, and expect to satisfy the function equation:

$$\pi^{-s/2}\,\Gamma\,(s / 2)\zeta(s) = \pi^{-(1-s)/2}\,\Gamma(1- s / 2)\;\zeta(1 - s)$$

In 1859, Riemann obtained an analytic formula for the number primes up to a pre-assigned limit. This formula is expressed in terms of the zeros of the zeta function. Riemann[1] introduced the function of complex variable t defined by:

$$\zeta\,(t) = 1/2\; s(s - 1)\;\pi^{-s/2}\,\Gamma(s / 2)\;\zeta(s)$$

and shows that $\zeta\,(t)$ is an even entire function of t whose zeros have imaginary part between $-i / 2$ and $i / 2$. Or simply: the nontrivial zeros of $\zeta(s)$ have real parts equal to ½ [1]

2. EARLY AND RECENT HISTORY

Sequence of events in the early history of mathematics and the analytic elements of the 17th, 18th and 19th centuries influenced Hilbert to announce that if the (RH) turn out to be true, it may lead to the exact formulation of the "Prime number Theorem" that may be the stepping stone to resolve the "Goldbach's and Twin Primes conjectures" at inaugural International Math Conference in 1900, and listed "Riemann Hypothesis, Goldbach's and Twin Primes" together with "The solvability of a Diophantine problem" as his 8th and 10th most important mathematical puzzle in the field of number theory.

(RH) re-appeared as a 2000 CMI Millennium problem together with:

Naiver–Stokes Equations (1830's) ....a problem from early physics
Poincare conjecture (1905) .... a modern topological problem
Hodge conjecture (1950) ....a modern topological problem
The Yang-Mills theory (1954) ....an important theory in physics
Birch & S-D conjecture (1960)................................. a modern study of an old Diophantine problem
P vs. NP problem (1971) .... a computer science problem

For following very good reasons:

(a) While Euler was investigating prime distribution via calculus and his zeta function in the 1730s; he also formulated three partial differential equations to describe Daniel Bernoulli's fluid flow problem --- a precursor of the 1830 Naiver-Stokes Equations.

(b) 40 years after the (RH), the Riemannian geometry (a broad version of differential geometry) was developed to study curve surfaces that motivated Henri Poincare to work on the foundations of modern topology --- a precursor of the 1905 Poincare conjecture, and subsequently, the 1950 Hodge conjecture.

(c) The same differential geometry was used by the early 20th century physicists to discover that quantum behavior of the physical world behaves very different from to the "classical" world. Typical of the quantum world is so called *wave-particle duality,* particles such as electrons behave sometimes as if they are point particles with a definite position, other times they are spread out like waves, their strange behavior is fundamental to the behavior of semiconductors in all our electronic devices, the nano-materials, and quantum computing [....] --- was a precursor of the 1954 Yang-Mills quantum theory.

(d) Hilbert's 10th problem and the modern theories of elliptic curves in the early 20th century --- was a precursor of the 1960 Birch and Swinnerton-Dyre conjecture .

(e) A major unsolved problem P vs. NP in computer science introduced by Stephen Cook in his 1971 computability theory [....] led Hugh Montgomery to discover a formula in 1972 about the spacing between the zeros of the zeta function and theorizing that his formula gave the spacing between the energy levels of what theoretical physicists called a complex quantum chaotic system. Followed by the idea of Allain Connes to pursue that quantum physics may lead to a proof of the (RH), and wrote down a system of equations that specified a hypothetical quantum chaotic system that has all the (primes) built in, and conjectured that this system has energy level corresponding to all the zeros of the zeta function that lie on the critical line. Consequently, the encryption community speculated that if (RH) can be resolved via quantum computing, then it may have ramifications in internet security --- it often based on factoring large prime numbers.

4. OBSERVATION AND REASONING

Since there is not yet a proposed strategy to handle this conundrum after 150 years, we can only rely on the observation of other reputable researchers together with the proof of Goldbach's conjecture in **(Exhibit A)** were credible --- as a counterexample to dispute the validity of the Riemann Hypothesis via reasoning:

(1) (RH) was built on an old abstraction (Euler's zeta function), his tentacles reached many other analytic elements, in attempt to prove another old abstraction (Gauss' density conjecture) --- but it was intelligible to the experts in the field, even to himself that he gave up after a few attempt. His colleague Flex Klein (1849-1925) commented that "we may never know why Riemann noticed the key link between Euler's zeta function and the primes that he found was a close connection between the density function $D_N = P(N) / N$ and solution to his $\zeta(s) = 0$ [3].

Moreover, in 1923, Carl Ludwig Siegel carried out a detailed study of Riemann's paper and reported "No part of Riemann's writing related to the zeta function is ready for publication; occasionally, one finds disconnected formulas on the same page; frequently just one side of equation has been written down; reminder estimates and investigations of convergence are invariably missing, even at essential point [4].

(2) In a major unsolved problem in computer science, P vs. NP introduced by Stephen Cook in his 1971 computability theory, Cook implied that if P = NP is proven; then it may lead to transform mathematics by allowing a computer to find a formal proof of any theorem which has a proof of a reasonable length, since formal proofs can be recognized in polynomial time. Example problems may include all the CMI problems.

(RH) is a CMI problem, it attempted to explain where every prime to infinity will occur by means of his abstract analytic formula $\zeta(s) = 0$. In essence, Stephen Cook questioned the validity of abstract modern mathematics behind the (RH) and asking if a fundamental problem as old as mathematics itself can be explained via a polynomial time procedure.

Hindsight, to uncover the pattern of the odd primes via a polynomial procedure elaborated in **(Appendix A-1)** in CHAPTER FOUR alone --- should constituted as a disproof of the (RH); and the notion that Riemann Hypothesis has connection with physics, cryptography is not foreordained facts. Undoubtedly, this proof will create havoc in modern mathematics and beyond.

External links are retrieved for information related to the Riemann Hypothesis only.

[1]: http://www.clay.org/millennium/riemannhypothesis.

[2]: G.H Hardy; E. M. Wright (2008) [1938]. An introduction to the theory of numbers (rev. by D. R Heath-Brown and J. H Silverman, 6th ed) Oxford University Press. ISBN 978-0-19-9211986-5 http://google.com/book?id = rey9wfSaJ9EC&dq

[3]: "The millennium problem" by Keith Devlin:
Riemann Hypothesis p. 19 – 62/ Navier-Stokes equations p. 132-154
Birch and S-D rank conjecture p. 190-211/ Hodge conjecture p. 213-228
Poincare conjecture p. 158-187/ Yang-Mills Theory p. 64-104/ P vs. NP p. 106 – 129

[4]: Carl Siegel, uber Riemann Nachlas zur analytischentheorie quellen und studien zur bechichte der mathematic, Astronmine und physic P.46

EXHIBIT 3   A DISPROOF OF THE (BSD) RANKING CONJECUTE

ABSTRACT: After proclaimed solutions of **(Exhibit A), (Exhibit C)** and **(Exhibit 2)** --- it became clear that the (BSD) conjecture is invalid due to: (a) it was based on obscure analytic number theory that has no real values, (b) the Diophantine equation that Birch and Swinnerton-Dyre was investigating can be resolved by a non-geometry-based polynomial procedure only.

1. THE PROBLEM

The official description of (BSD) conjecture in [2] was based on Wiles's understanding of the existing knowledge of elliptic curves, modularity and the 20th century deep theories of rational points on the curves but --- no one knows if they were true.

According to [3], with access of a powerful EDSAC computer and the 20th century modern theories of elliptic curves and functions; (BSD) calculated the zeta functions of certain elliptic curves of a Diophantine equation, their arithmetic aspects of calculation provided a detailed structure of the curves that resulted in the discovery of an analogue for an elliptic curve of the Tamagawa (1959) number of an algebraic group, but the algebraic aspects of calculation was extremely complicated in technique and intensive in computation. (BSD) indicated that they have detected certain relations between different invariants, but unable to prove these relations. In 1960, (BSD) conjectured that: *The elliptic curve will have infinite number of rational points if and only if L (E, 1) = 0.*

2. PRELUDE TO PROBLEM

Pythagorean right triangle formula is $X^2 + Y^2 = Z^2$ because $3^2 + 4^2 = 5^2$

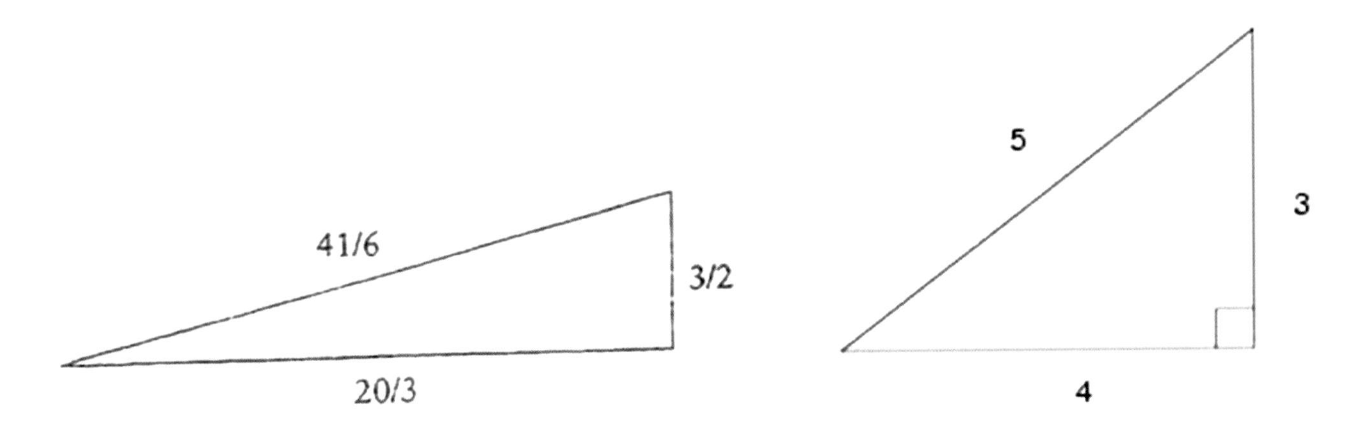

Modern researchers theorized that the Pythagorean right triangle with sides of 3, 4 and 5 has whole number solution of an area $d = (3 \times 4) / 2 = 6$ has an elliptic curve defined by algebraic equation of the general form as $y^2 = x^3 + ax + b$ because it's discriminate $\Delta = -16(27b^2 + 4a^3)$ is non-zero. But what about the right triangle with rational-number sides of 3/2, 20/3 and 41/6 that has whole number solution of an area $d = (3/2 \times 20/3) / 2 = 5$? Researchers say the right triangle with rational-number sides has an algebraic equation $y^2 = x^3 - d^2x$ turned out to have whole number solution due to its discriminate $\Delta = -16(-4d^2) = 64d^2$, which is non-zero, so the graph of this equation is an elliptic curve. Thus, to study elliptic curves, the only equations you need to look at are those of the general form of $y^2 = x^3 + ax + b$.

Diophantine problem of finding whole number d that are areas of right triangle with rational number-sides is equivalent to the problem of finding rational points (i.e., points whose coefficients are rational numbers) on certain elliptic curves. This is the problem (BSD) set out to investigate [2].

2. EARLY HISTORY

Pythagorean mathematics exists long before the Arithmetica (260 A.D) introduced a collection of problems giving numerical solutions of both determinate and indeterminate equations of Diophantine geometry. Diophantine studied integers, integers can be considered either in themselves or as solutions to equations. Diophantine also studied rational points on curves and algebraic varieties. In other words, he showed how to obtain infinitely many of rational points satisfying a system of equations by giving a procedure that can be made into an algebraic expression (basic algebraic geometry in pure algebraic forms). Diophantine was the first Greek mathematician who recognized fractions as numbers and allowed positive rational numbers for the coefficients and solutions. Diophantine equations are usually algebraic equations with integer coefficients, for which integer solutions are sought. Unfortunately, most of texts were lost or unexplained. Not very much happened in mathematics after the Greeks.

In the 17th century; the analytic geometry of Descartes, the elusive 1637 (FLT) and calculus of Newton, along with the analytic elements of the 18th and 19th centuries --- influenced David Hilbert to list "Riemann Hypothesis, Goldbach's and Twin primes conjectures" and "Solvability of a Diophantine equation" as his 8th and 10th problem respectively, and consequently, motivated:

(a) The 1901 work of rational points on curves and topology by Poincare
(b) The 1922 theory of elliptic curve by Mordell.
(c) The 1928 Mordell-Weil Theorem.
(d) The 1940s Hasse-Weil zeta functions in relation with Riemann's zeta function.
(e) The Shimura-Taniyama conjecture (1957) on elliptic curves in relation with modular forms [...]
(f) The work of Lang and Tate on Galois cohomology and Tamagawa numbers of motives, modular arithmetic in 1958 and 1959 respectively.
(g) The work on elliptic curves and others by the Tate-Shafarever group in the 1960(s)

Consequently, (BDS) and others believed elliptic curves was important and fundamental objects that link to number theory, geometry and topology; ranked the geometric properties of Diophantine problem in dimension one: Genus = 1 (elliptic curve), the detail structure is the (BSD)'s problem. Nevertheless, it is unnecessary to discuss elliptic curves any further since:

(h) The only reason (BSD)'s problem made the CMI listing was Andrew Wiles claimed a partial result on modularity in 1999 --- but no one knows if it is true.
(i) Nick Katz commented "A new idea is needed" at a 2001 Arizona winter school [3].
(j) **(Exhibit A), (Exhibit C)** proclaimed that problems in number theory were "arithmetic" in nature, and **(Exhibit 2)** disputed the Riemann Hypothesis.

5. RATIONAL POINTS ON HIGHER DIMENTIONAL VARIETIES

The official introduction of (BSD) conjecture in [2] stated that: Consider the conjecture of Euler from 1769 that $X^4 + Y^4 + Z^4 = t^4$ has no non-trivial solutions. But finding a curve of genus 1 on the surface and a point of infinite order on this curve, Elkies found the solution of $2682440^4 + 15365639^4 + 18796760^4 = 20615673^4$. Elkies argued that there are infinitely many solutions to Euler's equation.

*In fact, Euler investigated (FLT) extensively in the late 1730s without any success, but marked the "rebirth" of (FLT) as the beginning of modern number theory. Hindsight, Euler's conjecture is the variation of the (FLT) below:

"In about 1637, mathematician Pierre de Fermat (1601-1665) left a margin note in his copy of "Arthimetica" that then this indeterminate Diophantine equation $X^N + Y^N = Z^N$ has no integer solutions when N is an integer > 2"

*Actually, Euler's 1769 conjecture was correct and Elkies's argument will not stand --- this certainty is elaborated in the attached [(**Appendix 3A**)].

6. OBSERFATION

In a major unsolved problem in computer science P vs. NP --- introduced by Stephen Cook is his 1971 computability theory. Informally, "P" is a set of relatively easy problems, "NP" is a set of what seem to be very, very hard problems, so "P = NP" would imply that apparently hard problems actually have relatively easy solutions; author Stephen Cook implied that if P = NP is proven, then it may lead to transform mathematics by allowing a computer to a formal proof of any theorem which has a proof of a reasonable length since formal proofs can be recognized in polynomial time. Example problems may include all the CMI problems.

(BSD) conjecture is a CMI problem, but foremost, a Diophantine problem. When we speak of Diophantine equations today, we are speaking of polynomial equations, in which, integer solutions must be found. Polynomial is a mathematical expression that involves N's and N^2s and N's raise to other powers.

A new idea is discovered to illustrate that --- the original problem that (BSD) was investigating can be resolved by a non-geometry-based polynomial procedure via common factor. For details, please refer to the attached **[(Appendix 3B)]**

External links are retrieved for information related to BSD conjecture only:

[1]: http://www.clay.org/millennimn/birch and swinnerton-Dyreconjecture
[2]: "The Millennium Problems" by Keith Devlin / (BDS) conjecture p 190-211
[3]: BDS conjecture, a computational Approach wstein@math.washington.edu
[4]: P vs. NP in "The Millennium Problems" p.105- 129, p 110, p112, 215
[5]: Andrew Wiles, Modular elliptic curves and Fermat's Last Theorem, Ann Math 142 (1995) pp 442-551
[6]: http://en.wikipedia.org/wiki/Galois-cohomology
[7]: http://en.wikipedia.org/wiki/Galois-group

[(Appendix 3A)]

1. THE PROBLEM

Euler conjectured in 1769 that $X^4 + Y^4 + Z^4 = t^4$ has no non-trivial solutions. But finding a curve of genus 1 on the surface and a point of infinite order on this curve, Noan Elkies (Harvard) found the solution of $2682440^4 + 15365639^4 + 18796760^4 = 20615673^4$, and argued that there are infinitely many solutions to Euler's equation [1].

2. OBSERVATION AND SOLUTION

Please revisit **(Exhibit C),** actually, Euler's conjecture is the variation of the (FLT) --- it is equivalent to say that this indeterminate equation $X^N + Y^N + Z^N = t^N$ has no solutions when integer $N > 2$.

(BSD) conjecture came out of the 3rd century "Arithmetica", Diophantine recognized fractions as numbers and allowed positive rational numbers for the coefficients and solutions. Since we are dealing with positive integers, favor rigor (to mean polynomial). Based on our understanding of the (FLT), we believe the following **[Three studies]** adequately illustrated --- Euler's conjecture is correct and argument of Elkies will not stand:

[Study 1]: Euler did not constrain X, Y, Z. Therefore, there are two options:
$X = Y = Z$ or $X \neq Y \neq Z$

Let: $X = Y = Z = 2, N = 1$
Thus: $X^N + Y^N + Z^N = t^N$
Became: $2 + 2 + 2 = 6$ $t = 6$ (With common prime factor 2)

Variation: $X^N + Y^N = t^N - Z^N$ Let: $N = 1, 2, 3, 4 \ldots$
N = 1: $2 + 2 = 6 - 2$ $(2 + 2 = 4)$
N = 2: $2^2 + 2^2 \neq 6^2 - 2^2$ $(8 \neq 32)$
Because: $(2 + 2)^2 = 2^2 + (2 \times 2 \times 2) + 2^2 = 8$
Notice: $4 + 4 = 8$
Or: $2^2 + 2^2 = 2^3$

A polynomial equation of (a particular form) can be constructed to hold Euler' conjecture is correct that equation $(X^4 + Y^4 = t^4 - Z^4)$ has no non-trivial solution.

N = 4: $2^4 + 2^4 \neq 6^4 - 2^4$ $(16 + 16) \neq (1296 - 16)$
N = 5: $2_5 + 2^5 = 6^5 - 2^5$ $(32 + 32 \neq (7776 - 32)$
N = 10: $2^{10} + 2^{10} \neq 6^{10} - 2^{10}$ $(1024 + 1024) \neq (60466176 - 1024)$

[Study 2]: Let: $X \neq Y \neq Z$, $X = 1$, $Y = 2$, $Z = 3$ and $N = 1$:

$X^N + Y^N + Z^N = t^N$
Hence: $1 + 2 + 3 =$ 6 $t = 6$

Variation:	$X^N + Y^N = t^N - Z^N$.	Let: N = any positive integers
N = 1:	$1 + 2 = 6 - 3$	$(1 + 2 = 3)$
N = 2:	$1^2 + 2^2 \neq 6^2 - 3^2$	$(1 + 4 \neq 25)$
Because:	$(1 + 2)^2 = (6 - 3)^2$	
	$(3)^2 = (3)^2$	
Or:	$1^2 + (2 \times 1 \times 2) + 2^2 = 6^2 - (2 \times 6 \times 3) + 3^2$	$(9 = 9)$

A polynomial equation of a (particular form) can be constructed to hold Euler's equation $X^4 + Y^4 = t^4 - Z^4$ true that it has no non-trivial solutions.

N = 4:	$1^4 + 2^4 \neq 6^4 - 3^4$	$(1 + 16) \neq (1296 - 81)$
N = 5:	$1^5 + 2^5 \neq 6^5 - 3^5$	$(1 + 32) \neq (7776 - 533)$
N = 10:	$1^{10} + 2^{10} \neq 6^{10} - 3^{10}$	$(1 + 1024) \neq (60466176 - 1024)$

[Study 3]: To prove Elkies's calculation of $2682440^4 + 15365639^4 + 18796760^4 = 20615673^4$ was flawed:

In mathematics:

$\log A + \log B = \log A \times \log B$
$\log A - \log B = \log A / \log B$
$\text{Log } A^N = N \log A$.

Log 10 = 1, log 100 = 2, log 1000 = 3, log10000 = 4, so on and to forth

Assuming: $2682440^4 + 15365639^4 + 18796760^4 = 20615673^4$

Variation: $2682440^4 + 15365639^4 = 20615673^4 - 18796760^4$

$$\text{Log } 2682440^4 + \text{Log } 15365639^4 = \text{Log } 20615673^4 - \text{Log } 18796760^4$$
$$4 \text{ Log } 2682440 + 4 \text{ Log } 15365639 = 4 \text{ Log } 20615673 - 4 \text{ Log } 18796760$$
$$\text{Log } 2682440 + \text{Log } 15365639 = \text{Log } 20615673 - \text{Log } 18796760$$
$$6.428530017 + 7.186550541 \neq 7.314197517 - 7.27408299$$

3. THE CONCLUSION

Problems in number theory (or arithmetic) are best understood via study the "patterns of the numbers" and "elementary calculation techniques"

[(Appendix 3B)]

1. THE PROBLEM

A right triangle with rational-number sides of 3/2, 20/3 and 41/6 has a whole number solution of an area d = (20/3 × 3/2) / 2 = 5. The question is if this right triangle has infinitely many of whole number solutions of area d?

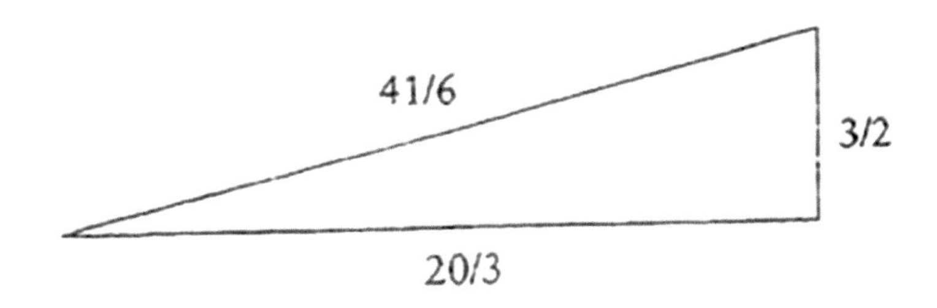

2. OBSERVATION AND SOLUTION

This right triangle has rational-number sides with height (3/2), base (20/3) and side (41/6). Notice the denominator of side (41/6) has common prime factor of 2 and 3. Therefore, the variation of the height and base can be expressed as: X = 3/2 = 9/6, Y = 20/3 = 40/6 respectively without change their real value. Notice that ---the right triangle with rational-number sides can be expressed by the formula of the Pythagorean right triangle:

$$X^2 + Y^2 = Z^2$$

$$(3/2)^2 + (20/3)^2 = (41/6)^2$$

Variation: $9/6)^2 + (40/6)^2 = (41/6)^2$

Or: $(9)^2 + (40)^2 = (41)^2$ (81 + 1600 = 1681)

A polynomial algorithm (of a particular form) is formulated so that its solution to one (n) can be adapted to solve all the other (n)s:

$$(9n)^2 + (40n)^2 = (41n)^2 \ldots\ldots\ldots\ldots(1) \text{ Let: } n = 1, 2, 3 \ldots$$

n =1: $(9)^2 + (40)^2 = (41)^2$

With an area: $d = (40 \times 9) / 2 = 360 / 2 = 180$

n =2: $(9 \times 2)^2 + (40 \times 2)^2 = (41 \times 2)^2$

$(18)^2 + (80)^2 = (82)^2$ (324 + 6400 = 6724)

With an area: $d = (80 \times 18) / 2 = 1440 / 2 = 720$

n =3: $(9 \times 3)^2 + (40 \times 3)^2 = (41 \times 3)^2$

$(27)^2 + (120)^2 = (123)^2$ (729 +14400 = 15129)

With an area: $d = (120 \times 27) / 2 = 3240 / 2 = 1620$

n =10: $(9 \times 10)^2 + (40 \times 10)^2 = (41 \times 10)^2$

$(90)^2 + (400)^2 = (410)^2$ $(8100 + 160000 = 168100)$

With an area: $d = (400 \times 90) / 2 = 36000 / 2 = 18000$

3. THE CONCLUSION

If a right triangle $X^2 + Y^2 = Z^2$ with rational-number sides has a whole number solution of an area d; then X, Y and Z have a common prime factor.

EXHIBIT 4 A REVISIT OF THE POINCARE CONJECTURE (PC)

ABSTRACT: Poincare's conjecture (PC) was supposedly resolved in 2003 by Grigio Perelman --- siding his certain technique illustrated the power of his differential geometric approach to question in topology was consistent with the current knowledge of Ricci flow (a certain partial differential equation), and accepted by the experts in 2006, but Perelman declined the honor and prize in 2010.

Nevertheless, we disputed the validity of the (PC) based on our understanding that "topology" is arithmetic in nature --- that has no connection with calculus of any variations; topology only has multiple connections with: a rigorous infinite topological field mapped out based on proving the Goldbach's principle (1+1) and straightedge and compass construction, Therefore, a revisit of (**EXHIBIT A, B, C, D, E, F, G**) and early history leading to the Poincare conjecture will be necessary.

1. THE PROBLEM

In 1905, Poincare conjectured that "*Every simply connected, closed 3-manifold is a homeomorphic to the 3-sphere*. Informally, (PC) is an abstract algebraic topological theorem based on Poincare's intuitive understanding the theory: (a) that modern algebraic geometry has multiple conceptual connections with such diverse fields as complex analysis, topology and number theory, (b) of the 20th century Riemannian geometry (elliptic or spherical geometry), a very broad version of differential geometry originated with the vision of Bernard Riemann expressed in his inaugural lecture --- on the Hypotheses on which Geometry is based [6].

2. EARLY AND RECENT HISOTRY

ALGEBRAIC GEOMETRY --- *"Algebraic geometry in pure algebraic form"* --- was introduced in the 3rd century "Arithmetica". Diophantine studied rational points on curves (elliptic) and algebraic varieties. In other words, he showed how to obtain infinitely many of the rational points satisfying a system of equations by giving a procedure that can be made into as algebraic expression, but most of the texts were lost or unexplained. Some of the roots of algebraic geometry date back to the work of Hellenistic Greeks (450 BC). The Delain problem --- "doubling the cube" and other related problems such as; "trisecting the angle", "polygons", "squaring the circle" --- they are also known as straightedge and compass problems (or topological problems). Not very much happened to mathematics after the "Arithmetica".

In the 17th century, Galileo (1564-1642) was the father of "observational astronomy" and "modern science", a polymath in the field of mathematics, physics, engineering and natural philosophy:

In mathematics --- Galileo applied the standard passed down from ancient Greeks and Fibonacci (1170-1250); but superseded later by the algebraic methods of Descartes.

* Rene' Descartes (1596-1630) made a fundamental discovery that by --- assuming by restrict ourselves to the "straightedge and compass" in geometry, it is impossible to construct segments of every length. If we begin with a segment of length 1, say, we can only construct a segment of another length if it can be expressed using integers, addition, subtraction, multiplication, division and square roots (as the golden ratio can). Thus, one strategy to prove that a geometric problem is impossible (not constructible) --- is

to show that the length of some segment in the final figure cannot be written in this way. But doing so rigorously required the nascent field of algebra [….]

Descartes introduced the analytic geometry --- primarily to study algebraic curves to reformulate the classic works on conic and cube; using Descartes' approach, the geometric and logical arguments favored by the ancient Greeks for solving geometric problems could be replaced by doing algebra "*algebraic geometry to extend the mathematical objects to multidimensional and Non-Euclidean spaces*". Descartes also studied <u>tangent</u> line. * During the same period, Pierre de Fermat (1601- 1665) studied prime numbers, tangent lines to curves and independently developed analytic geometry *to study the properties of algebraic curves (those defined in Diophantine geometry), which is the manifestation of solutions of system of polynomial equations*. Fermat was famous for his elusive (1637 FLT), and his limited prime number 5, 7, 17, 257, 65537.

Desargues (1591-1661) introduced the projective geometry to study geometric properties that are invariant under projective transformations; mainly to deal with those properties of geometric figures that are not altered by projecting their image onto another surface. Desargues also contributed in Blaise Pascal's Theorem.

It is worth noting that Pascal, Desargues and others in the 17th century argued against the use of algebraic and analytic methods in geometry; they also studied curves from a purely geometrical point of view: the analog of the Greek ruler and compass construction --- from their perspective without any success.

In physics --- Galileo's theoretical and experimental work on the motions of bodies, along with the work of Kepler (1571-1630) and Descartes was a precursor of the classical mechanics developed by Newton (1643-1727) and Leibniz (1646-1716). Initially, calculus was introduced to study physical problems, historically, the first method of doing so was by infinitesimals, but infinitesimals do not satisfy the Archimedean property.

CALCULUS --- is still to some extent an active area of research today even after centuries of improvement. According to [3] --- the ancient period introduced some of the ideas that led to integral calculus, but does seen to have developed these ideas in a rigorous and systematic way. Calculations of volumes and areas; one goal of integral calculus, can be found in Egyptian Moscow Pepurus (1820 BC), but the formulas are mere instructions, with no indication as to method, and some of them lack major components. From the age of Greek mathematics, Eudoxus (400 BC) used the method of exhaustion, which foreshadows the concept of the limit to calculate areas and volumes, while Archimedes (250 BC) developed this idea further, inventing heuristics which resemble the methods of integral calculus. The method of exhaustion was later reinvented in China by (Liu Hui) in the 3rd century AD in order to find area of a circle.

In the 5th century, Zu Chong-Zhi established a method that would later be called Cavalier's principle to find the volume of a sphere [….]

In the 18th century, Daniel Bernoulli (1700-1782) was the first --- to accept analytic geometry (Descartes) mainly because it supplied with concrete quantitative tools needed to analyze his physical fluid motion problem via infinitesimal calculus, and his colleague Leonhard Euler (1707-1783) contributed greatly; and during the same period, Euler became interested in number theory after his friend Goldbach (1690-1764) introduced Fermat's unelaborated work (including the FLT) to him and adapted infinitesimal calculus to investigate prime distribution via his zeta function. Although Euler was unsuccessful in all fronts but his investigation of Fermat's unelaborated work marked the "rebirth" of the (FLT) as the beginning of modern number theory. Consequently, Euler's analytic approach influenced:

(a) The 1791 observation of prime distribution in a sufficiently large number by Carl F. Gauss (1777-1855), together with the Legendre's (1752-1833) prime distribution conjecture --- became the Gauss' density conjecture due to their similarity.
(b) In geometry --- Gauss proved in theory that a (17-sided polygon) was constructible via "Straightedge and compass construction" led us to believe geometry has connections with prime numbers but no one was able to do so (including himself).
(c) In algebraic geometry --- researchers theorized that it has multiple conceptual connections with such diverse fields as complex analysis, topology and number theory; nevertheless, it is still in the uncharted area of mathematics today.

TOPOLOGY --- To the ancient Greeks, topology was the mathematical study of shapes and spaces --- it concerned with the basic properties of space, such as connectedness, continuity and boundary. The earliest topological problems dated back to the work of Hellenistic Greeks (450 BC). The Delian problem --- "doubling the cube", and other related problems such as; "trisecting the angle", "polygons" and "squaring the circle" --- also known as "Ruler and compass problems.

The motivating insight behind "topology" was that some of these problems depend not on the exact shape of the objects involved, but rather on the way they put together. For example, the square and the circle have many properties in common: they are both one dimensional object (from a topological point of view) and both separate the plane into two parts; inside and outside parts, not much more can be said.

(a) In geometry, Leibniz (1646-1716) defined tangent line is a line through a pair of infinitely close points on the curve, one of the most fundamental notions in differential geometry.
(b) While Euler was investigating prime distribution via calculus, he also published his 1736 paper on the bridges of Kongsberg --- which classified as topological.
(c) Evaraiste Galois (1811-1832) introduced his theory to study roots of polynomials and researchers say that Galois group theory has been used to prove that "doubling the cube" and "trisecting the angle" was "insoluble" and characterizing the regular polygon was "constructible" [....]
In 1829, Lobachevsky published the non-Euclidean "Hyperbolic geometry" after researchers failed to understand the complexity of Euclid's parallel postulate; his version of parallel postulate led to the "elliptic geometry" and claimed if one takes the parallel postulate as given, then the result is "Euclidean geometry"--- but his paper was rejected.
(d) During the 1850s --- in connections with the basic problems of surveying and geodesy; Carl F. Gauss initiated the field of differential geometry. Using differential calculus, he characterized the intrinsic properties of curves and surfaces. For instance, he showed that the intrinsic curvature of a cylinder is the same as that of a plane, as can be seen by cutting a cylinder along its axis and flattening, but not same as that of a (sphere), which can not be flattened without distortion.
(e) In 1853, Riemann published "Habilitation" on the foundation of geometry" and the "Hypothesis" in 1859. Moreover; the 1865 Riemann-Roch theorem implied cube curve $y = x^3$ of a sphere is similar to the parabola curve $y = x^2$, therefore, the cube curve must have a singularity "at infinity"; as all its

points in the affine space are regular. Thus, many of the properties of the algebraic varieties and all the topological properties depend on the behavior "at infinity" --- which implied the projective space played a fundamental role in algebraic geometry; consequently, led to the development of Riemann surfaces in connection with elliptic curves.

(f) In 1874, the work of Georg Canter led the researchers to believe modern topology depends on the ideas of the set theory. Canter considered point sets in Euclidean space as part of his study of Fourier series.

(g) In about 1887, a lost paper to prove the "doubling the cube" was "unsolvable" by Pierre Wantzel (1837) was discovered; moreover, Wantzel applied the similar method to dispute the solvability of "trisecting the angle" and "polygons". In 1882, Ferdinand von Lindermann proved that the "squaring the circle" was "unsolvable".

* *That is why modern topologists do not mention and the Math Journals do not accept these "problems of antiquity" anymore. Nevertheless, study of topology went on in advanced study.*

(h) In 1895, Poincare introduced the concepts of homotopy and homology in his book "Analysis Situs" and marked the beginning of modern topology. Moreover, Henri Poincare was a polymath in modern mathematics and physics:

*In mathematics --- Poincare's work of rational points on algebraic curves and topology in 1901 together with the Riemannian geometry (elliptic or spherical geometry) introduced in the early 20th century to study surface of the curves (a broad and abstract generalization of the differential geometry of surfaces in R^3) that can be applied to study of differentiable manifolds of higher dimensions and spurred the development of algebraic and differential topology. In 1905, Poincare conjectured that "Every simply connected, closed 3-manifold is a homeomorphic to the 3-sphere"

*In physics --- Poincare (1854-1912), Einstein (1879-1955) and Lorentz (1858-1947) had studied the symmetries in Maxwell's equations (1860s), and it was their combined work that led to discover the Lorentz symmetry --- which lies at the heart of relativity, and consequently, led **Hermann Wekl's** (1885-1955) to discover a new symmetry of electromagnetism, now known as gauge symmetry [9] .

3. OBSERVATION

(1) There is no evident to suggest that finding infinitely many of the rational points on the curves was possible, and algebraic geometry is still in uncharted area of mathematics.

(2) (PC) did not grow out of blue; Poincare generalized that a certain three-dimensional shape **(no name was mentioned)** is equivalent to a three-dimensional sphere [2] --- based on his understanding of the analytic geometry, topology of a sphere and the Riemannian geometry that:

* Behind topology was that some geometric problems depend not on the exact shape of the objects involved, but rather on the way they put it together.

* Calculus and topological transformation have everything to do with (infinitely small).

* Topological transformation is that two points will not be destroyed, and remained (infinitely close) after continues deformation, shrinking, stretching, folding, but not tearing, thus do not depend on notions such as straight line, circles, cubes, or on measurement of lengths, areas, volumes or angles.

** Poincare certainly echoed the "doubling the cube", and "trisecting the angle" which were proven as "unsolvable" by Pierre Wantzel in 1832 and the "squaring the circle" which was proven by Ferdinand von Lindermann as "unsolvable" in 1882 --- but their proofs have no real values since* ***(Exhibit A, B, D, E, F and G)*** *illustrated that they were "arithmetic" in nature.*

According to [5], a sphere is a geometric object in 3-dimensional space that is the surface of a ball. Like a circle in a 2-dimensional space, a sphere is mathematically defined as the set points that are all have the same distance r from a given point. The surface, volume of a sphere were defined by Archimedes; its curves, the 3-dimentional equations, geometric properties and topology on a sphere in relation with circle were studied in the 19th century via the analytic geometry [....] In topology, an n-sphere is defined as a space homeomorophic to the boundary of an (n+1) ball, thus, it is homeomorphic to the Euclidean n-sphere, but lacking its metric. Researchers generalized: *A 0-sphere is a pair of points with the discrete topology. *A 1-sphere is a circle (up to homeomorphism); thus any knot is a 1-sphere. *A 2-sphere is an ordinary sphere (up to homeomorphism); thus any spheroid is a 2-sphere. *A n-sphere is a compact topological manifold without boundary [sic]

** It is unnecessary to discuss topology any further since we have already claimed the subject of topology has no connection with calculus.*

(3) In 2000, CMI listed --- the P vs. NP introduced by Stephen Cook in 1971 as one of their millennium problems; and Cook implied if P = NP is proven, then it may lead to transform mathematics by allowing a computer to find a formal proof of any theorem which has a proof of a reasonable length, since formal proofs can be recognized in polynomial time. Example problems may include all the CMI problems.

** The 1905 Poincare's Conjecture is a CMI problem; in essence, Stephen Cook was questioning the validity of abstract modern mathematics behind the (PC) and asking if Poincare's topological problem can be resolved via a polynomial time procedure. Polynomial is a mathematical expression that involves N's and N^2s and N's raised to other powers.*

(4) In a short answer of topology [4] by Robert Bruner (Wayne State) --- he stated "In topology, any continues change which can be continuously undone is allowed. So a circle is same as a triangle or a square, because you just "pull on" parts of the circle to make corners and then straighten the sides, to change a circle into a square. Then you just "smooth it out" to turn back to a circle; these two processes are continuous in the sense that during each of them, nearby points at the start is still nearby at the end. The circle is not same as the shape of a (figure ∞), because although you can squash the middle of a circle together to make it into a (figure ∞) continuously, when you try to undo it, you have to break the connection in the middle and this is discontinuous: points that are all near the center of the (figure ∞) end up split into two batches, on opposite sides of the circle, far apart [....]

(5) Keith Devlin (Stanford) stated in his 2002 book entitled "The Millennium Problems" that --- For a century now, mathematicians have built new abstractions on top of the old ones, every their new step taking further away from the original subjects on which, ultimately, and we must base all our standing. It is not so much that mathematician does new things; rather, the object considered became more abstract, abstractions from abstractions [2].

Furthermore, Devlin stated --- Poincare conjecture arose by accident, as a result of mistake Poincare make (and quickly noticed) right at the start of his investigation of this (new geometry). Much of the interest in topology focuses on 3 or 4 dimensional objects, and Poincare's mistake was to assume that a fairly obvious fact about 2-dimentional objects **(no name mentioned)** was also true for analogous objects of 3 or more dimensions. This 2-dimentional topology is sometimes called "rubber sheet geometry"

Indeed, from our perspective, this 2-dimensional "rubber sheet geometry" that **(no name mentioned)** was the "squaring the circle" --- the only 2-dimensional topological problem left by the ancient Greeks --- that was declared as "unsolvable" in 1882 by Ferdinand von Lindermann --- siding π was not constructible [...] and later by Courant/Ribbons [7].

*Nevertheless, we have a technique to handle the **π** via "straightedge and compass construction" Please revisit **(Exhibit A, B)** and **(Exhibit D, E, F)**, then the "squaring the circle" in **(Exhibit G)**.*

External links are retrieved for general information of "Topology" only:

[1]: The official description of the "Poincare conjecture" by CMI
[2]: "The millennium problems"/ Poincare conjecture / Keith Devlin p. 7, 157-187, 215
[3]: http://en.wikiprdia.org/wiki/calculus
[4]: "What is Topology, any way?" math.wayne.edu/-rrb/topology.html
[5]: http://en.wikipedia.org/wiki/Sphere
[6]: http:/ maths.tcd.ie/pub/HisMath/People/Riemann/Geom
[7]: "What is Mathematics?" 15th edition by Courant and Ribbon
[8]: Hazewinkel, Michiel, ed .(2001), "topology, general", Encyclopedia of mathematics Springer, ISBN 078-1-55608-010-4
[9]: Gerardus 't Hooft, editor. 50 years of Yang-Mills theory. World science, NJ 2005

EXHIBIT 5 TO DISPUTE THE VALIDITY OF THE HODGE CONJECTURE (HC)

ABSTRACT: Researchers say Hodge's conjecture is a major unsolved topological problem grew out of the projective space of algebraic geometry, derived by calculus variations (differentiation, integration). The official description of the (HC) in [1] was beyond our understanding, but no one knows if it is true anyway.

Mathematics is a sequential topic. Each branch is a connected whole, after disputed the Poincare conjecture in **(Exhibit 4)** --- it became clear that the Hodge conjecture is invalid because "topology" has no connection with calculus of any fashion; Topology only has multiple connections with: (a) A rigorous infinite topological field constructed after proving the Goldbach's principle, (b) the technique of ruler and compass construction.

Nevertheless, Hodge conjecture warrants further discussion because --- it is characteristic of the "doubling the cube" --- the only three-dimensional unsolvable topological problems left by the ancient Greeks --- which we have a certain understanding from a purely 'arithmetic" point of view via "straightedge and compass construction".

1. THE PROBLEM

Researchers say --- after De Rham (1903-1990) identified the De Rham cohomology groups as topological invariants and provided ways to investigate the shapes of complicated objects in his 1931 theorem; Hodge attempted to measure degrees of connectivity using algebraic constructs such as homology and homotogy groups grow out of algebraic geometry introduced in a 1895 book "Analysis situs" by Henri Poincare.

Informally, Hodge's problem was the result of his attempt to enrich the description of De Rham cohomology to include extra structure which was present in the case of complex algebraic varieties. *In 1950, Hodge conjectured that "On a projective non-singular algebraic variety over* ***C*** *(complex plain) any Hodge class is a rational liner combination of classes cl(Z) of algebraic circles".*

2. ALGEBRAIC GEOMETRY AND CURVES

Mathematicians have: (a) puzzled with the paradoxes of the "pattern of the primes" and "concept of infinite", (b) questioned how to find and describe the intersection of algebraic curves in the early development of algebraic geometry --- long before the "Element" (350 B.C) introduced the 1st proof of the infinitude of primes by abstract reasoning and devoted a part to primes, divisibility, topics unambiguously belong to number theory. Euclid also referenced tangent to a circle introduced Euclidean geometry and considered the line and the circle were the fundamental curves in geometry.

Some of the roots of algebraic geometry dated back to the work of Hellenistic Greeks (450 BC), the Delian problems --- the "doubling the cube" and other related geometry problems such as; "trisecting the angle", "polygons" and "squaring the circle" --- they are also known as "straightedge and compass problems" (or topological problems):

"Algebraic geometry in pure algebraic form" --- was initially introduced in the 3rd century "Arithmetica". Diophantine studied rational points on curves (elliptic) and algebraic varieties. In other words, he showed how to obtain infinitely many of the rational points satisfying a system of equations by giving a procedure that can be made into as algebraic expression, but most texts were lost or unexplained.

In the 17th century, Rene' Descartes (1596-1630) made a fundamental observation (discovery): Assuming by restrict ourselves to the "straightedge and compass" in geometry, it is impossible to construct segments of every length. If we begin with a segment of length 1, say, we can only construct a segment of another length if it can be expressed using integers, addition, subtraction, multiplication, division and square roots (as the golden ratio can). Thus, one strategy to prove that a geometric problem is impossible (not constructible) is to show that the length of some segment in the final figure cannot be written in this way. But required the nascent field of algebra [....]

Descartes introduced the analytic geometry primarily to study algebraic curves to reformulate the classic works on conic and cube; using Descartes' approach, the geometric and logical arguments favored by the ancient Greeks for solving geometric problems could be replaced by doing algebra *"algebraic geometry to extend the mathematical objects to multidimensional and Non-Euclidean spaces"*. Descartes also studied <u>tangent</u> line. * It was during the same period; Pierre de Fermat (1601- 1665) studied prime numbers, tangent lines to curves, and independently developed analytic geometry to study the properties of algebraic curves *(those defined in Diophantine geometry)*, which is the manifestation of solutions of system of polynomial equations. Fermat was famous for his elusive 1637 (FLT).

Desargues (1591-1661) introduced the projective geometry to study geometric properties that are invariant under projective transformations; mainly to deal with those properties of geometric figures that are not altered by projecting their image onto another surface. Desargues also contributed in Blaise Pascal's Theorem.

**It is worth noting that Pascal, Desargues and others argued against the use of algebraic and analytic methods in geometry; they also studied curves from a purely geometrical point of view: the analog of the Greek ruler and compass construction.*

In physics --- Galileo's theoretical and experimental work on the motions of bodies, along with the work of Kepler (1571-1630) and Descartes was a precursor of the classical mechanics developed by Newton (1643-1727). Initially, calculus was introduced to study physical problems, historically, the first method of doing so was by infinitesimals, but infinitesimals do not satisfy the Archimedean property. Leibniz (1646-1716) invented calculus independently,

In the 18th century:

Taylor series (1715) was introduced to represent a function as an infinite sum of terms that are calculated from the values of the function's derivatives at a single point. Eventually, it became common practice to approximate a function by finite number of terms of its Taylor series.

Physicist/mathematician Daniel Bernoulli (1700-1782) accepted the 17th century analytic geometry mainly because it supplied with concrete quantitative tools needed to analyze his physical fluid motion problem via infinitesimal calculus, and his colleague Leonhard Euler (1707-1783) contributed greatly; and during the same period, Euler became interested in number theory after his friend Goldbach (1690-1764) introduced Fermat's unelaborated work (including the FLT) to him and adapted infinitesimal calculus to investigate prime distribution via his zeta function. Although Euler was unsuccessful in all fronts but his work on (FLT) marked the "rebirth" of the (FLT) as the beginning of modern number theory. Evidently:

(a) In number theory --- Gauss's (1777-1855) 1791 observation of prime distribution in a sufficiently large N, together with Legendre's (1752-1833) prime distribution conjecture became the Gauss' density conjecture; and subsequently, most algebraic character of coordinate geometry was subsumed by the calculus of infinitesimals of Lagrange and Euler.

(b) In geometry --- the mathematical community accepted that --- geometry has connection with Fermat's prime numbers after Gauss proved in theory that a 17-sided polygon was constructible via "Straightedge and compass construction", but no one was able to do so. Moreover, Leibniz defined tangent line is a line through a pair of infinitely close points on the curve, one of the most fundamental notions in differential geometry. .

(c) In physics --- Pierre-Simon (1749-1827) introduced the Laplace equations --- a 2nd order of partial differential equation. Moreover, he summarized and extended (celestial mechanics) and translated the geometry study of classical mechanics to one based on calculus and opened up a broader range of problems in mathematical analysis, fluid dynamics, astronomy and physics.

(d) In algebraic geometry --- researchers theorized that the classic algebraic geometry studying properties of the sets of polynomial equations [7] and:

* It has multiple conceptual connections with such diverse fields as complex analysis, topology and number theory. The most studied classes of algebraic varieties are --- plain algebraic curves, which include lines, circle, cubic-curves like elliptic curves.

* It studied a point of the plane belongs to an algebraic curve if its coordinates satisfy a given polynomial equation. Basic questions involve the study of singular points, the inflection points and the points at infinity. More advanced questions involved the topology of the curve and relations between the curves given by different equations

*The projective space of algebraic geometry in (HC) may influenced by the projective geometry of Desargues --- that studied the geometric properties that are invariant under projective transformations.

3. HODGE CONJECTURE

In a 2002 book entitled "The millennium Problems" [2], Keith Devlin (Stanford) re- introduced the (HC) as "*Every harmonic differential form (of a certain type) on a non-singular projective algebraic variety is a rational combination of cohomology classes of algebraic cycle*". Devlin described the (HC) by --- assuming we could formulate the (HC) by starting with integrals over generalized paths in algebraic varieties, deforming the paths leave the values of such integrals unchanged, so you can think of the integrals as being defined on the class of paths.

Hodge's theory is that if certain of those integrals are zero, then there is a path in that class that can be described by polynomial. Every harmonic differential form refers to a solution to a certain partial different equation (Laplace equation) arrived from study functions of complex numbers, algebraic variety is complex algebraic variety, referring to algebraic equations.

Hodge proposed that you can start with any finite collection of equations, the algebraic variety consists of all points that solve all the equations in the system, and a complex variety is projective if the solutions of define equations depend only on the ration of the numbers involved, and non-singular if the surface is smooth.

Cohomology can be viewed as a method of assigning algebraic invariants to a topological space that has a more refined algebraic structure than homology. A rational combination of cohomology classes of algebraic cycle is referring to rise of the class of abstract objects from the concept of a complex analytic function. In Hodge's case; calculus is not done in real or complex number, it is on much more general kinds of varieties (algebraic equations). It deals with calculus developed on a non-singular projective variety and claims that a certain kinds of abstract objects arise, when you start with a certain kind of variety and does some differentials calculus on it.

3. TOPOLOGY

To the ancient Greeks, topology is the mathematical study of shapes and spaces --- it concerned with the basic properties of space, such as connectedness, continuity and boundary. The earliest topological problems dated back to the work of Hellenistic Greeks (450 BC). The Delian problem --- the "doubling the cube" and other related problems such as --- "trisecting the angle", "polygon" and "squaring the circle" --- also known as "Ruler and compass problems". The motivating insight behind topology is that some geometric problems depend not on the exact shape of the objects involved, but rather on the way they put together. For example, the square and the circle have many properties in common: they are both one dimensional object (from a topological point of view) and both separate the plane into two parts; inside and outside parts, no more can be said.

(a) In the 18th century, while Euler was investigating prime distribution, he also published his 1736 paper on the bridges of Kongsberg --- which classified as topological.

(b) In the early 19th century --- Evaraiste Galois (1811-1832) introduced his deep theory to study roots of polynomials; and it was said that Galois group theory had been used to prove "doubling the cube and trisecting the angle" were "insoluble", and characterizing the regular polygon was "constructible" [....]

(c) In the 1850s --- in connections with the basic problems of surveying and geodesy; Carl Gauss initiated the field of differential geometry via differential calculus, he characterized the intrinsic properties of curves and surfaces. For instance, Gauss showed that the intrinsic curvature of a cylinder is the same as that of a plane, as can be seen by cutting a cylinder along its axis and flattening, but not same as that of a sphere, which can not be flattened without distortion.

(d) In 1853, Euler's student Riemann published his "Habilitation" on the foundation of geometry, and the 1859 "Hypothesis". In complex analysis, Riemann introduced: (i) the Riemann to parabola surfaces,

(ii) the 1865 Riemann-Roch theorem, which implied the cube curve of a sphere $y = x^3$ is similar curve $y = x^2$, so the cubic curve must be have a singularity, which must be "at infinity". Thus many of the properties of the algebraic varieties, including all the topological properties depends on the behavior "at infinity" [....], and theorized the projective space plays a fundamental role in algebraic geometry.

(e) Influenced by the 17th century algebraic method of Descartes, Pierre Wantzel proved that the "doubling the cube" was "unsolvable" in his 1837 paper, he also applied the similar method to dispute the solvability of "trisecting the angle" and "polygons", unfortunately, his paper did not re-appear 50 years later. Moreover, in 1882, Ferdinand von Lindermann proved that the "squaring the circle" was "unsolvable"--- siding π was not constructible.

* *That is why mathematical Journals do not accept and modern topologists do not mention those problems of antiquity --- anymore. Nonetheless, topology was still an important subject for advanced study.*

(f) In 1895, Poincare published his book entitled "Analysis situs" to introduce the concepts of homotopy and homology --- and marked the beginning of modern topology.

(g) In early 20th century, Riemannian geometry was introduced to study the surface of sphere (elliptic geometry or spherical geometry); it is a broad abstract generalization of the differential geometry of surfaces in R^3. Researchers say the development of R-geometry resulted in synthesis of diverse results concerning the geometry of surfaces and the behavior of geodesics on them, with techniques that can be applied to study of differentiable manifolds of higher dimensions. It enabled Einstein's general relatively theory, made profound impact on group theory, representation theory as well as analysis, and spurred the development of algebraic and differential topology. In 1905, Poincare conjecture was coined.

(h) In 1950, Hodge conjecture (HC) was coined, but no one knows if it is true. W.V.D Hodge (1903-1975) was also known for Hodge star operator, Hodge bundle and Hodge theory [9].

5. OBSERVATION AND REASONIMG

(1) In 2000, CMI listed a computer science problem P vs. NP introduced by Stephen Cook in 1971 as one of their millennium problem, Cook implied that if P = NP is proven, then it may lead to transform mathematics by allowing a computer to find a formal proof of any theorem which has a proof of a reasonable length, since formal proofs can be recognized in polynomial time. Example problems may include all the CMI problems.

* *Hodge Conjecture is a CMI problem. In essence, Stephen Cook was questioning the abstract modern mathematics behind the 1950 Hodge conjecture and asking if it can be resolved by a certain polynomial time procedure.*

(2) In a 2002 book entitled "The Millennium Problems" by Keith Devlin (Stanford), he stated that --- For a century now, mathematicians have built new abstractions on top of the old ones, every their new step taking further away from the original subjects on which, ultimately, and we must base all our

standing. It is not so much that mathematician does new things; rather, the object considered became more abstract, abstractions from abstractions [2].

Indeed, historically, mathematicians used existing theory to identify useful overarching principle that can guide the development of new theories --- the only other topological problem left by the ancient Greeks with a possible projective nature was the "doubling the cube", moreover, it is characteristic of the "Hodge conjecture" from our perspective:

** "Doubling the Cube" asked to increase dimension of a cube to double the volume of a given cube --- ancient Greeks could never have proved it based on their knowledge of the "ruler and compass construction", but the brightest knew it was possible.*

** Roughly, Hodge asked to what extent you can approximate the shape of a given object by gluing together simple geometric building blocks to increase dimension. Although Hodge could not prove it, but intuitively he knew this classes of mathematical objects is not only projective but also have infinite solutions --- free of calculus [2]*

Modern topologists do not mention the "doubling the cube" anymore because it was proven as "unsolvable by Pierre Wantzel in 1832 and also by Courant and Ribbons in [7].

*Nevertheless, with **(chart E-1 in Exhibit B),** we can look at the "doubling the Cube" from a purely arithmetic point of view --- via straightedge and compass construction, please revisit **(Exhibit F)** for details.*

External links are retrieved for general information of topology and Hodge conjecture only. They are not linked to the proclaimed proof:

[1]: http://www.clay.org/millennium/Hodge conjecture
[2]: "The millennium problems"/ Hodge conjecture by: Keith Devlin p.214 - 228
[3]: Kelley, Hohn L (1975) Gerneral Topology. Sringer-Verlag. ISBN 0-387-90125-6
[4]: "What is Mathematics?" 15th edition, (1973) by courant and Robbins
[5]: Berger, Marcel (2000) Riemann geometry during the 2nd halt of the 20th century. Americain Math Society, ISBN0-8218-2052-4
[6]: http://en.wikipedia.org/wiki/ricmannspace
[7]: http://en.wikipedia.org/wiki/algebraicgeometry
[8]: http://en.wikipedia.org/wiki/w.v.dhodge
[9]: http://en.wikipedia.org/wiki/Hodge-star-operator
[10]: http://en.wikipedia.org/wiki/differenetialgeometry
[11]: http://en.wikipedia.org/wiki/differenetialcalculus

EXHIBIT 6 A DISCUSSION OF THE YANG-MILLS QUANTUM THEORY (YMQT)

ABSTRACT: *God created the universe and natural numbers, everything else is men made. Infinite universe and natural numbers; macroscopic in distance, microscopic up close.* Informally, Einstein's classical theory describes the universe in macroscopic scale and quantum theory describes the universe in microscopic scale. The (YMQT) challenges us to find a rigorous grand unified theory **(GUT)** proposed by physicists during the 1970s to explain behavior of the classical and quantum world we live in.

The objective here is to claim: (i) the (GUT) was only a conjecture, (ii) uncovering the pattern of the odd primes via a polynomial procedure to resolve the Goldbach's conjecture elaborated in **(Exhibit A)** is the only possible "mathematical theory" to explain behavior of (macroscopic and microscopic world) we live in because --- it inspired to map out an infinite mathematical field **(Exhibit B)** that not only disputed the topological problems arose in (QFT) analogously, but also unified all the pre-Newtonian mathematics to perform rigorous calculation and outperforming any computer both in speed and range (however large or small).

1. PRELUDE TO PROBLEM

Plato theorized in "Timaeus" (350 BC) that fire, water, earth and air were all aggregates of tiny solids [....]. Galileo theorized in the 17th century that --- the mathematics passed down by ancient Greeks was the language with which God has written the universe. In the 20th century, physicists thought they had a solid grasp on what made up "matter" until two important discoveries in 1932 --- the neutron split atoms in two and the framework for the fundamental "Standard model". The (YMQT) predicted correctly for all important standard model other than Gravity.

The official CMI description of the (YMQT) is elaborated in [1]; but our understanding was based on a 2011 short conversation between Professor Michel Murray (University Adelaide), Alan Carey and Peter Bouwknegt (Australian national U) [3] as follows:

Physicists discovered in the early 20th century that at a very short distance, such as the size of an atom or smaller, the world behaves very differently to the "classical" world we are used to. Typical of the quantum world is so called *wave-particle duality*, particles such as electrons behave sometimes as if they are point particles with a definite position, other times they are spread out like waves. Moreover, their strange behavior is not just of theoretical interest, since it underpins much of our modern technology. It is the fundamental to the behavior of semiconductors in all our electronic devices, the nano-materials, and the rise of quantum computing.

Quantum theory is fundamental. It must govern not only just "microscopic small" but also the "classical" realm. That means physicists and mathematicians have to develop methods not just for understanding new quantum phenomena, but also replacing classical theories by their quantum analogues --- this is the process of *Quantization;* when we have a finite number of degrees of freedom; such as for a finite collection of particles.

Although the quantum behavior is often counter-intuitive, we have a well-developed mathematical machinery to handle this quantization called *quantum mechanics* --- this is well understood physically and mathematically; but when we move to study the *electric and magnetic field* where we have an infinite number of degrees of freedom, the situation is much more complicated. With the development of so-called *quantum field theory* (QFT), a quantum theory for fields, physics has made progress that mathematically we do not understand.

Many fields theories fall into a class called *gauge field theories*, where a particular collection of symmetries (called the gauge group), acts on the fields and particles. In the case that these symmetries all commute, so-called *abelian gauge theories*, we have a reasonable understanding of the *quantization*. This includes the case of the electromagnetic field, quantum electrodynamics --- which the theory makes impressively accurate predictions. Historically, the first example of a *non-abelian theory* that arose is the theory of *the electro-weak interaction*, which requires a mechanism to make the predicted particles massive as we observed them in nature. This involves the *Higgs boson*, which is being researched for with the large Hadron collier (LHC) at CERN --- because the Higgs mechanism is classical and carries over to the quantum theory under the quantization process.

2. THE PROBLEM

The 1954 (YMQT) is *a non-abelian gauge theory* --- it describes all the forces of nature other than Gravity, and predicted correctly for all important standard model in physics, which we expect to describe quarks and strong force that binds the nucleus and powers Sun. Here, the conflict between the *classical theory* and *quantum theory* began:

The classical theory predicts mass-less particles and long-range forces. The quantum field theory (QFT) has to match the real world with short range forces and massive particles. Moreover, physicists expect various mathematical properties such as *"mass gap" and "asymptotic freedom"* to explain the non-existence of mass-less particles in observations of the strong interactions. As these properties are not visible in the classic theory and arise only in the quantum theory, understanding them means a rigorous approach to the (YMQT) is needed --- but currently, we do not have the mathematics to do this, although various approximations and simplifications can be done which suggest the quantum theory has the required properties.

In 2000, CMI selected the 1954 (YMQT) as their prize problem and challenged anyone to establish a rigorous unified mathematical theory (GUT) purposed by theoretical physicists of the 1970s to explain behavior of the classical and quantum world we live in to prove the existence of the "mass gap" that is; the none-existence of mass-less particles.

3. DISCUSSION

Calculus was initially introduced to study physical problems. Historically, the first method of doing so was by infinitesimals, but infinitesimals do not satisfy the Archimedean property. As we know: (i) calculus has two major branches, differential calculus (concerning rate of change and slopes of curves) and integral calculus (concerning accumulation of quantities and the areas under curves), (ii) calculus played major role in the saga of non-rigor modern mathematics that influenced Hilbert to list the "Riemann Hypothesis, Goldbach's and Twin Primes conjecture" and the "Solvability of a Diophantine equation" as his 8th and 10th problem in 1900; and calculus played vital roles in modern physics:

(1) Historically, Daniel Bernoulli (1700-1782) was the first mathematician/physicist to accept the 17th century Descartes's analytic geometry mainly because it supplied with concrete quantitative tools needed to analyze his physical fluid motion problem via infinitesimal calculus; and his colleague Leonhard Euler (1707-1783) contributed three partial differential equations to describe Bernoulli's fluid flow problem after his death.

Pierre-Simon Laplace (1749-1827) introduced a 2nd order of partial differential equation; summarized and extended (celestial mechanics) and translated the geometry study of classical mechanics to one based on calculus and opened up a broader range of problems in mathematical analysis, fluid dynamics, astronomy and others.

In 1840s, William Hamilton's (1805-1865) mechanics [....] and its close connection with geometry served as a link between classical and quantum mechanics

In 1865, James Maxwell (1831-1879) published his book entitled "A Dynamical Theory of the Electromagnetic Field".

In 1901, Faraday (1791-1867) established the basic concept of electromagnetic field which led Wilhelm Rontgen (1845-1923) to produce and detect electromagnetic radiation in a wavelength range (X-ray) that earned him the inaugural Nobel Prize.

In 1902, Lorentz (1858-19470 share the Nobel Prize with Zeeman (1865-1943) because --- Poincare (1854-1912), Einstein (1879-1955) and Lorentz (1858-1947) studied the symmetries in Maxwell's equations, and it was their combined work that led to discover the Lorentz symmetry --- which lies at the heart of relativity. Moreover, in mathematics; Poincare's concepts of homotopy and homology in his 1895 book "Analysis Situs" marked the beginning of modern topology, his work of rational points on algebraic curves and topology in 1901 together with the Riemannian geometry (a broad and abstract generalization of the differential geometry of surfaces in R^3) enabled to study surface of the curves that --- can be applied to study of differentiable manifolds of higher dimensions and spurred the development of algebraic and differential topology. Poincare was known for his 1905 topological conjecture.

In 1921, Einstein's (1879-1955) Nobel Prize for his relativity theory was influenced by Minkowski's theory of relativity, along with Hilbert's axioms, space [...] together with Hermann Weyl's (1855-1955) work in number theory, topology, symmetry, general relativity in theoretical physics and the work of Arthur Compton. Weyl's discovered a new symmetry of electromagnetism --- now known as gauge symmetry, and also held the chair at Gottingen in 1930s.

In 1922, Bohr's (1855-1962) won the Prize for his contribution of the understanding on atomic structure and quantum theory.

In 1932, Heissenberg (1901-1976) won the Prize for theory of quantum mechanics.

In 1935, James Chadwick (1891-1974) won the Prize for discovery of neutron.

In 1950, Hodge's topological conjecture was coined, and researchers say Hodge star-operation played role in Yang-Mills equations [7][8].

In 1954, the (YMQT) was accepted based on --- non-rigor mathematical and physical theories of different researchers in different time for completely different reasons.

In 1957, C N Yang and T D Lee shared the Prize for their interpreted results of particle decay experiments at Brookhaven's Cosmotron particle accelerator and their work of the elementary particles [...]

(2) In (QFT), theoretical physicists say: (i) when we have a finite number of degrees of freedom; such as for a finite collection of particles, although the quantum behavior is often counter-intuitive, we have a well-developed mathematical machinery to handle this quantization called *quantum mechanics,* (ii) the non-abelian theory that arose is the theory of the electro-weak interaction requires a mechanism to make the predicted particles *massive* as we observed them in nature --- *this involves the Higgs boson.*

In 1964, mathematician/physicist Steve Hawking (1942-2018) argued and bet that Higgs boson would never be found.

In 1965, Feynman, Tomonaga and Schwinger jointly won the Prizes for Feynman's work in the path integral formulation of quantum mechanics and [....]

In 1971, a major problem in computer science P vs. NP introduced by Stephen Cook in his computability theory implied that if P = NP, then the Yang-Mills theory may have a polynomial time solution (to mean rigor).

In 1979, Hawking elected Lucasian Professor of mathematics at U of Cambridge, made a transition in his approach to physics based more on intuition and speculation rather than insist on rigorous mathematical proofs. MOREOVER, Weinberg, Glashow and Salam shared the 1979 Nobel Prize for formulating and establishing two standard models of fundamental interactions and cosmology [....] in early 1970s.

In 1981, Hawking told Kip Thorne that "I would rather be right than rigorous".

In 1984, Rubbia and Simon van der Meer shared the Prize for discovery of the field particles W and Z, communicators of the weak interaction.

In 1992, Georges Charpak won the Prize "for his invention and development of particle detectors and the multi-wire proportional chamber."

In 1994, Shull (1915-2001) and Brockhouse (1918-2003) shared the Prize for their pioneering work and the development of neutron scattering techniques [...]

In 1999, the Prize was awarded jointly to Hooft and Veltman --- for elucidating the quantum structure of electro-weak interactions in physics *(It was their work that put the unifications theory proposed by Weinberg, Glashow and Salam on the map)*

(3) In 2000, CMI listed the (YMQT) as their prize problem to challenge the 21st century mathematician/ physicist or anyone to find a grand unified Theory (GUT) --- the extensions of standard models formulated by Weinberg and Salem in the 1970s

In 2011, Higgs boson was confirmed by the ATLAS and CMS based on the collisions in the Large Hadron Collider (LHC) at CERN; and Hawking conceded the bet.

In 2013, Higgs and Englert won the Nobel Prize for their theoretical predication.

In 2015, the Prize went to Kajita and MacDonald for their discovery of neutrino oscillations, which shows that neutrinos have mass.

In 2016, the Prize was divided to Kosterlitz, Haldane and Thouless for their theoretical discoveries of topological phase transitions and topological phases of matter.

In 2017, Kip Thorne and Barish shared the Prize with Weiss --- for their contribution to the LIGO detector and the observation of gravitational waves.

Nevertheless, in 2018, two articles published in Newscientist.com/issue/3182 based on the data obtained 5 years after researches at CERN; many questions respect to the Higgs boson and quantum mechanics surfaced:

"Higgs works but still baffles" by Richard Webb (chief editor)

We were expecting it --- but it only deepened the enigma. Experiments at the Large Hadron Colluder at CERN near Geneva, found that the Higgs boson really does give all other fundamental particles mass.

The new measurements, by the ATLAS and CMS collaborations, have seen the Higgs, discovered at CERN in 2012, being produced along with both a top quark and its antimatter equivalent, a top antiquark. It follows the discovery by both groups of the Higgs boson decaying directly into **tau** leptons, heavier cousins of the electron (arxiv.org/abs/1708.00373). Both measurements are based on five years of data.

Interactions between the Higgs and the top quark are particularly key, as the top is the most massive fundamental particle. At 172 gigaeletronvolts, it dwarfs even the Higg's own mass of 125 GeV. So what ever Higg does, it does more with the top "it drives why the universe is like it is," says Fabio Cerutti of ATLAS. But it's really puzzling: we don't have to understand why the top mass is so heavy." The results suggest the interaction between the Higgs and the top quark is just as predicted by the standard model, our best description of how particles and forces interact (arxiv.org/abs/ 1806.00425). The tau lepton results are similarly in line. Together, they are the first direct indication that interactions with Higgs boson explain why fermions, the class of particles that make up matter have the masses they do. Understanding the Higgs mechanism could help explain the stability and structure of matter around us.

Subtle differences in the masses of up and down quarks explain why neutrons within the atomic neutrons are heavier than protons, and so why stable atoms can form. Similarly, the mass of the electron sets the size of atoms, and so determine how chemistry works. The top quark measurements are another bittersweet triumph for the standard model, says Jon Butterworth of the ATLAS team. "It is one of the benchmark processes we knew we had to see as part of the health check of the standard model," he says "And that health is disgustingly good." Disgusting, because we know the standard model is incomplete. It leaves open essential questions on the make-up of dark matter, which is more than 2 / 3 of all matter, and why matter apparently dominates antimatter in the universe. "Every time we see something that agrees the standard model, that means we don't know the answers, he says "Any deviation would be a clue, and we want clues.

"Beyond Quantum"

By Claran Lee (a quantum researcher at University College London) commented that there are fundamental questions that quantum mechanics just can not answer and theoretical predictions that violate its premises [...]

(4) The same differential geometry and calculus used by physicist/mathematician of the 20th century in their development of (QFT) also make close predictions about geometry and topology in low dimensions; in particular path integrals, but mathematically, we have no idea what is this path integral, except in very elementary cases it is as if we are back to the pre-Newtonian world, certain calculations can be done with certain tricks.

Analogously, there were calculations in geometry and topology that can be done non-rigorously using methods developed by physicists in (QFT); *this led researchers to believe there is a 21st century math theory with powerful techniques to be discovered.*

(5) In a major unsolved computer science problem P vs. NP introduced by Stephen Cook in his 1971 paper "The complexity of theorem-proving procedures"; Stephen Cook implied if P = NP is proven, then it may allowing a computer to find a formal proof of any theorem which has a proof of a reasonable length, since formal proofs can be recognized in polynomial time. Example problems may include all the CMI problems.

The (YMQT) is a CMI problem --- it studied the "natural universe" via modern abstract mathematics. In essence, Stephen Cook questioned the validity of non-rigor modern mathematical and physical theories behind this problem, and asking if a rigorous <u>mathematical theory</u> can be found via a polynomial time procedure to explain the classical and quantum world we live in.

* Hindsight, solving the principle problem in number theory via a polynomial procedure inspired to construct a rigorous infinite mathematical field elaborated in **(Exhibit A)** and **(Exhibit B)** is the only math theory that --- not only represented as a rigorous infinite topological field to resolve the topological problems in **(Exhibit F)** and **(Exhibit G)** in order to dispute the validity of two CMI topological problems in **(Exhibit 4 and 5)**, but also unified the pre-Newtonian mathematics to perform rigorous calculation that outperforming any computer both in speed and range **<u>however large or small --- which expressed the behavior of classical and quantum world</u>** we live in.

(4) In 2021, Science [6] published a report by a group of researchers about the topological system in relation with the work of Quantum physics by Wekl (1885-1955)

Yes, Weyl worked on number theory, topology, gauge symmetry and general relativity, and the generalization of the concept of band topology in [6] and close predictions of topology in low dimension --- may consistent with the work of Weyl; nevertheless, we can only infer this report has no real value since we have claimed that problems in number theory and topology were "arithmetic" in nature --- actually, nothing occurs contrary to nature except the impossible.

Moreover, according to [5] --- Weinberg's (1933-2021) work was incomplete because: (i) Weinberg conjectured, but could not prove that his unified theory [...], (ii) it was just a theory of leptons --- electrons, muons and their neutrinos, but left out the world of strongly interacting hadrons [...]

External links are retrieved for general information related to the Yang-Mill theory only:

[1]: http://www.clay.org/millennium/Yang-mills theory
[2]: "The millennium problem" by Dr. Keith Devlin / p 64- 104 Yang-Mills Theory
[3]: The Conversation 12 / 8, 2011. Millennium prize: the Yang-Mills existence and Mass gap problem by: Michel Murray, Alan Carey and Peter Bouwknegt
[4]: http://wikipedia.org/wiki/path-integral/mass-gap
[5]: Nature /Vol 596 / 8-12-2021/ Page 183 / Steven Weinberg
[6]: "Linked Weyl surfaces and Weyl arcs in photonic metamaterials" published by "Science" page 572-576/Vol 373 / Issue 6534 / 7-30-2021
[7]: http://wikipedia.org/wiki/hodgestar-operator
[8]: Gerardus 't Hooft, editor. 50 years of Yang-Mills theory. World science, NJ 2005

EXHIBIT 7 A DISCUSSION OF THE NAIVER-STOKES EQUATIONS

ABSTRACT: The 1850 N-S equations is a CMI problem grew out of: (a) the 17th century Newton's 2nd law, (b) the 18th century Euler's three partial differential equations to describe Daniel Bernoulli's fluid flow problem, (c) enhanced in the 19th century by themselves and others. Although this improved version contributed greatly in wide range of practical uses --- but researchers have no idea how to find a formula to prove it exists to "infinity" because they are not polynomial.

After proclaimed the solution of the CMI problems elaborated in **(Exhibit 1 to 6)** --- it became very clear that the N-S equations should be an interest in pure mathematics. The objective here is to revisit the early history and circumstance that led to Euler's three partial differential equations; then to discuss this CMI problem from our point view --- via reasoning.

1. EARLY HISTORY

Galileo (1564-1642) was the father of "observational astronomy" and "modern science", a polymath in mathematics, physics, engineering and natural philosophy:

In mathematics --- he applied the standard passed down from ancient Greeks and Fibonacci (1170-1250); but superseded later by the algebraic methods of Descartes.

Rene' Descartes (1596-1630) introduced analytic geometry primarily interested to study algebraic reformulation of classic works on conic and cube. Using Descartes' approach, the geometric and logical arguments favored by the ancient Greeks for solving geometric problems could be replaced by doing algebra (algebraic geometry extended mathematical objects to multidimensional and Non-Euclidean spaces), he also wrote a book about the tangent line problem.

It was during the same period, Pierre de Fermat (1601-1665) introduced analytic geometry independently to study the properties of algebraic curves (those defined in Diophantine geometry), which is the manifestation of solutions of system of polynomial equations. Fermat also using Diophantine approximates equality to find maxima for functions and tangent lines to curves.

Nevertheless, Desargues (1591-1661), Pascal (1623-1662) and many other 17th century mathematicians argued against the use of algebraic and analytic methods in geometry.

In physics --- Galileo's theoretical and experimental work on the motions of bodies, along with Descartes and Kepler (1571-1630) were the precursor of the classical mechanics developed by Newton.

Initially, Newton (1643-1727) developed calculus to study physical problems; calculus is a collection of techniques for manipulating infinitesimals, capable of to approximate a polynomial series, so infinitesimals do not satisfy the Archimedean property.

Historically, Daniel Bernoulli (1700-1782) was the 1st physicist /mathematician to accept the 17th century analytic geometry of Descartes for it supplied with concrete quantitative tools needed to analyze his physical fluid motion problem via infinitesimal calculus and his colleague Leonhard Euler contributed three partial differential equations to describe Daniel Bernoulli's fluid flow. It was during the same period, Euler: (a) Became interested of number theory after his friend Christian Goldbach (1690-1764) introduced him the unelaborated work of Fermat (including the FLT), (b) Adapted infinitesimal calculus and his zeta function to analyze prime distribution.

Although Euler was unsuccessful in all fronts; but marked the 1637 (FLT) as the beginning of modern number theory. Evidently, his analytic approach via modern calculus paved the way of modern study of mathematics and physics:

In modern mathematics, in order to put calculus in a more solid footing: (a) Mathematician Cauchy (1789-1857), a pioneer in early math analysis formalized the concept of infinite by defined "continuity", (b) Wrierstrass (1815-1897) introduced elliptic curves in normal form as $Y^3 = x^2 + ax + b$, and gave the definition of limit (eliminated infinitesimal). Infinitesimals were replaced by very small numbers, and the infinitely small behavior of the function is found by taking the limiting behavior for smaller and smaller numbers. So, calculus is a collection of techniques for manipulating numbers and certain limits. Eventually, it became common to base calculus on limit and paved the way for modern study of number theory.

In late 18th century, Pierre-Simon (1749-1827) introduced the Laplace equations --- a 2nd order of partial differential equation, moreover, he summarized and extended (celestial mechanics) and translated the geometry study of classical mechanics to one based on calculus and opened up a broader range of problems in mathematical analysis, fluid dynamics, astronomy and physics.

* In fluid dynamics, physicist Claude-Louis Navier (1785-1836) improved Euler's three partial differential equations, and George Stokes (1819-1903) enhanced

2. PRELUDE TO PROBLEM

The official description by Fefferman in [1] was from a professional point view, but to prove one of his statements in (A), (B), (C) and (D) in page 2 are still unanswered.

Keith Devlin described the N-S Equations for non-professionals in [2] as --- Daniel Bernoulli regarded motions of the fluid as made up of infinitesimally small discrete regions, infinitesimally close together, and each of these regions can be handled using calculus and published his "Hydrodynamics" in 1738.

Let's begin with Euler's equations for fluid motion, the equations that govern flow in a (hypothetical) frictionless fluid that extends to infinity in all directions.

We assume that each point *P* = (x, y, z) in the fluid is subject to a force that varies with time. We can specify the force at P at time t by giving its values in each of the three axial directions: f_x (x, y, z, t), f_y (x, y, z, t), f_z (x, y, z, t).

Let *P* (x, y, z, t) be the pressure in the fluid at the point *P* at time t; then the motion of the fluid at point *P* at time t can be specified by giving it velocity in the three axial directions:

u_x (x, y, z, t) be the velocity of the fluid at *P* in the x-direction
u_y (x, y, z, t) be the velocity of the fluid at *P* in the y-direction
u_z (x, y, z, t) be the velocity of the fluid at *P* in the z -direction

We assume that the fluid is incompressible. That is when a force is applied to it, it may flow in some direction but it cannot be compressed; nor can be expand. This is expressed by the following equation:

$$\text{(a)} \qquad \frac{du_x}{dx} + \frac{du_y}{dy} + \frac{du_z}{dz} = 0$$

The problem assumes that we know how the fluid is moving at start, i,e., when $t = 0$. That is, we know u_x (x, y, z, 0), u_y (x, y, z, 0), and u_z (x, y, z, 0) --- as functions of x, y, and z. Moreover, these initial functions are assumed well-behaved ones. (Exactly what this means is technical, but we don't require a definition in order to obtain an overall understanding of the problem). The precise formulation of the restriction is relevant to one of the four statements in [1] --- (A), (B), (C) and (D) on page 2.

Applying Newton's law f = ma to each *P* point in the fluid, when combined with the incompressibility equation (a), Euler produced and left the following unsolved equations (b), (c) and (d):

$$\text{(b)} \qquad \frac{du_x}{dt} + u_x \frac{du_x}{dx} + u_y \frac{du_x}{dy} + u_z \frac{du_x}{dz} = f_x(x,y,z,t) \frac{dp}{dx}$$

$$\text{(c)} \qquad \frac{du_y}{dt} + u_x \frac{du_y}{dx} + u_y \frac{du_y}{dy} + u_z \frac{du_y}{dz} = f_y(x,y,z,t) \frac{dp}{dy}$$

$$\text{(d)} \qquad \frac{du_z}{dt} + u_x \frac{du_z}{dx} + u_y \frac{du_z}{dy} + u_z \frac{du_z}{dz} = f_z(x,y,z,t) \frac{dp}{dz}$$

3. THE PROBLEM

The N-S Equations involved the Bernoulli's fluid flow, Euler's three partial differential equations and Newton's law f = ma --- with the assumption that the stress is the sum of a diffusing viscous term (proportional to the gradient of velocity) and a pressure term, hence describing viscous flow. Navier and Stokes asked if Euler's three equations could be extended to "infinity" after their following improvement:

(1) To allow for viscosity, a positive constant *v* was introduced to measure the frictional force within the fluid, and added an additional force --- the viscous force:

$$\text{(e)} \qquad v \left[\frac{d^2u_x}{dx^2} + \frac{d^2u_x}{dy^2} + \frac{d^2u_x}{dz^2} \right]$$

This term (e) is to be added to the right hand side of equation (b), with entirely similar terms (with u_x replaced by u_y and u_z, respectively) added to equations (c) and (d). With the assumption that there has to be a pattern to describe the motion of fluid, the differential calculus was used to describe and analyze motion change.

Here, the notation $\frac{d^2u_x}{dx^2}$ denotes the 2nd partial derivative, obtained by 1st differentiating with respect to x and then differentiating the result again with respect to *x*, i.e.,

$$\frac{d^2u_x}{dx^2} = \frac{d}{dx} \times \frac{du_x}{dx}$$

with analogous definitions in the *y* and *z* cases.

Differential calculus --- is a technique to describe and analyze motion and change, not just any motion of change; there has to be a pattern that describes its occurrence. Therefore, differential calculus is a collection of techniques to manipulate patterns. The aim of differentiation is to obtain the rate of change of some changing quantity --- in order to do this, the value or position or path of that quantity has to be given by an appropriate formula. Differentiation then acts upon this formula to produce another formula that gives the rate of change. Differentiation is a process for turning formulas into other formulas.

(2) In the 20th century, new development in modern mathematics and physics [....] led physicists to write the Navier-Stokes equations more compactly as:

$$\text{(f)} \qquad \frac{du}{dt_x} + (u. \nabla)u = f - \text{grad } p + v\,\Delta\, u \qquad \text{div } u = 0$$

Here, f and u are *vector functions* and the symbols/terms Δ ∇, grad and div denote operations of *vector calculus.*

Vector function --- the tangent vector field (an assignment of a tangent vector) can be used to represent a direction of fluid travels, and developed a notation and a method to handle directional motion in a simpler fashion. Whereas a number only has quantity, a vector has both direction and quantity.

Vector calculus is the method you get when you develop calculus for vector and vector functions instead of number-variable and number-variable functions --- vector calculus is a mathematical study of tangent vector field by the physicists in the early 19th century.

The improved version of (f) by the 20th century physics works in wide range of practical uses, but no one knows if a certain "formula" can be found to satisfy the equations to "infinity".

4. OBSERVATION AND REASONING

(1) While Euler was investigating prime distribution via calculus and his zeta function in the 1735s; he also formulated three partial differential equations to describe Daniel Bernoulli's fluid flow problem (the precursor of the N-S equations). Moreover, in about 1742, after Euler announced the Goldbach (his friend)'s observation of prime distribution was accurate but he could not prove it mathematically; this principle problem in pure mathematics transformed into a problem of analytic number theory.

** Both the 1742 Goldbach's conjecture and Naiver-Stokes equations (1830) involved calculus and looking for **(a certain formula)** to prove their problems exist to **"infinity"**.*

(2) In P vs. NP, a major unsolved problem in computer science introduced by Stephen Cook's 1971 computability theory; Dr. Cook implied if P = NP is proven, then it may lead to transform mathematics by allowing a c, omputer to find a formal proof of any theorem which has a proof of a reasonable length, since formal proofs can be recognized in polynomial time. Example problems may include all the CMI problems.

** Naiver-Stokes Equation is a CMI millennium problem; in essence, Stephen Cook questioned the non-rigor physical mathematics and asking if a certain polynomial time algorithm can be formulated?*

(3) Please revisit **(Exhibit A)** and **(Exhibit B)** --- had the 20th century physicists known the fact that a "formula" to hold the **(GC),** the principle problem in pure mathematics true to "infinity" --- via an unprecedented polynomial formula only, then they would not have questioned if their improved compact Navier-Stokes equations (f) can extend to "Infinity"

(f) $$\frac{du}{dt_x} + (u. \nabla)\, u = f - \text{grad } p + v\, \Delta\, u \qquad \text{div } u = 0$$

Nevertheless, this improved (f) works successfully in fluid dynamics, heat transferring, airplane design and wide range of practical uses in science and technology beyond our imagination --- because calculus was invented to study physical problems; capable of to approximate a polynomial series only, and a close approximation is good enough in science and technology.

External links are retrieved for information related to Navier –Stokes equations only:

[1]: http://www.clay.org/millennium/Naiver-Stokes equation
[2]: "The millennium problems" by Keith Devlin / N-S Equations p. 131-157, p.215
[3]: "The millennium problems" by Keith Devlin / P vs. NP/ p. 106-129

THREE

OTHER INTRACTABLE PROBLEM IN MATHEMATICS

****

The proclaimed topics in CHAPTER TWO adequately illustrated that --- mathematical knowledge is not a collection of isolated facts. Each branch is a connected whole; linked to other branches, but ultimately, they are all connected to the roots of mathematics: the pattern of the primes. Moreover:

****

* With the proclaimed solution of the Goldbach's conjecture in **(EXHIBIT A),** we can look at the elusive "Twin primes conjecture" observed by Alphonse de Polignac in 1849 from our perspective --- since they both deal with odd prime numbers.

* With the proclaimed proof of the Fermat's Last Theorem and (BSD) ranking conjecture elaborated in **(EXHIBIT C) and (Exhibit 3)** respectively, now, we are in position to resolve the 1993 Beal conjecture (observed by Andrew Beal in 1993).

* With the proclaimed solution and technique of "Trisecting the angle" and "Heptagon") elaborated in **(EXHIBIT D and E) ---** to construct "a 17-sided polygon" via "Straightedge and Compass construction" is within reach.

****

For complete details, please refer to (Exhibit 8), (Exhibit 9) and (Exhibit 10) in the next few pages

****

EXHIBIT 8 A PROOF OF THE TWIN PRIMES CONJECTURE

ABSTRACT: After uncovered the pattern of odd primes to resolve the Goldbach's conjecture elaborated in (Exhibit A); it became clear that "Twin Primes" is best understood to study together with the Goldbach's conjecture --- since they both deal with infinitude of odd primes. The objective here is to illustrate that the "Twin Primes conjecture" can be resolved by a polynomial procedure, but via an axiomatic method only. Polynomial is a mathematical expression that involves N's and N^2 s and N's raise to other powers.

1. THE PROBLEM

Twin primes is defined as a pair of odd primes with a form of [p p+2]. Since 2 is an even prime, therefore, the gap of (2, 3) is not considered as [p p+2]. It was officially conjectured by Paul Stackel (1862-1919) that there are infinitely many pairs of "Twin primes".

2. EARLY AND RECENT HISTORY

"Twin primes" was first observed by Alphonse de Polignac (1826-1890) in 1849. He extended [p p+2] to the idea that there are infinitely many pairs for any possible finite gaps, such as; [p p+4], [p p+6] --- not just [p p+2]. It is also a mathematical true in pure mathematics. So "Twin primes" could be called as "Polignac conjecture".

After Hilbert listed "Riemann Hypothesis, Goldbach's and Twin Primes conjectures" as his 8th problem in 1900; "Twin primes" was accepted as an open problem in analytic number theory. "Twin primes" is similar to the 1742 Goldbach's conjecture; both deal with odd primes but impossible to exhaust all the numbers to reach a conclusion. Reputable mathematician Brun (1915), Hardy-Littlewood (1930s), Bombieri (1960s) and others conveyed their intuitive understanding of both the (GC) and Twin primes. .

The 20th century John Friedlander, Henryk Iwaniec and the 21st century Dan Goldston, Janos Pintz and Cem Yildirim all attempted and failed to prove that there are infinitely many pair of primes with some finite gaps. In 2013, the work of Iwaniec influenced Dr. Yitang Zhang's paper entitled: Bounded gaps between Primes [5]:

$$\liminf_{n \to \infty} (P_{n+1} - P_n) < 7 \times 10^7$$

was accepted and published by "Annals" [5] as a milestone achievement because his major ingredient was a stronger version of the Bombieri -Vinogradov Theorem (1960s).

3. OBSERVATION

(a) With respect to the Bombieri-Vinogradov Theorem --- it was a major result of analytic number theory, concerning the distribution of primes in arithmetic progressions averaged over a range of moduli --- refined based on Mark Barban's result in 1961 and Vinogradov published on a related topic, the density hypothesis in 1965. This result is a major application of the large Sieve method [....]

(b) With respect to the technique of Iwaniec --- Henryk Iwaniec together with John Friedlander proved in 1997 that there are infinitely many of prime numbers of the form of $a^2 + b^4$. The result of this proof [...] led the researchers in the field to conclude --- Iwaniec made deep contribution in the field of analytic number theory, mainly in modular form on GL(2) and Sieve methods.

***** ***Nevertheless, they were only abstractions from abstractions --- based on their intuitive understanding of the analytic number theory that has no real values.***

(c) With respect to the Zhang's proof of $\liminf_{n \to \infty} (P_{n+1} - P_n) < 7 \times 10^7$

***** ***(7×10^7) is a sufficiently large number (a finite number), Zhang did not prove the conjecture; nonetheless, his work was accepted because it was consistent with the existing knowledge, but what about to prove [p, p +4] and [p, p+6] < 7×10^7?***

4. REASONING

Polignac conjectured that Twin Primes as [p, p+2] and it can be extended to [p, p+4], [p, p+6] --- since proving the Goldbach's principle (1+1) in **(Exhibit A)** and Twin Primes both deal with odd primes, Twin primes is best understood as follows:

[Study1]: With the odd primes in **(Appendix A-1)** and the technique of ascending and descending order of odd numbers (vertically) in the range of any even number $N \geq 6$:

N = 8: In the upper half, we have three primes --- 3, 5, 7

1 + 7
3 + 5
------
5 + 3
7 + 1

Number pairs comply with **[p, p+2]:** (3, 5), (5, 7)
[p, p+4]: none
(1+1): (8 = 3 +5)

N= 32: In the upper half, we have ten primes --- 3, 5, 7, 11, 13, 17, 19, 23, 29, 31.

1 + 31
3 + 29
5 + 27
7 + 25
9 + 23
11 + 21
13 + 19
15 + 17

↓

Number pairs comply with **[p, p+2]:** (3, 5), (5, 7), (11, 13), (17, 19), (29, 31)
[p, p+4]: (3, 13), (5, 17), (7, 19), (13, 23), (19, 29)
[p, p+6]: (3, 19), (5, 23), (17, 31)
(1+1): (32 = 3 + 29), (32 = 13 + 19),

N = 100 In the upper half, we have:

1 + 99	24 primes: 3, 5, 7, 11, 13, 17, 19, 23, 29, 31, 37, 41, 47, 53, 59, 61, 67, 71, 73, 79, 83, 89, 97
3 + 97	
5 + 95	
7 + 93	Comply with **[p, p+2]:** (3, 5), (5, 7), (11, 13), (17, 19), (29, 31), (41, 43), (59, 61), (71, 73)
9 + 91	
11 + 89	
13 + 87	**[p, p+4]:** (3, 13), (5, 17), (7, 19), (11, 23), (13, 29), (17, 31), (19, 37), (23, 41), (29, 43), (31, 47). (37, 53), (41, 59), (43, 61), (47, 67), (53, 71), (59, 73), (61, 79), (67, 79), (71, 89), (73, 97)
17 + 83	
19 + 81	
21 + 79	
23 + 77	
25 + 75	**[p, p+6]:** (3, 19), (5, 23), (7, 29), (11, 31), (13, 37), (17, 41), (19, 43), (23, 47), (31, 59), (37, 61), (41, 67), (43, 71), (47, 73), (53, 79), (59, 83), (61, 89), (67, 97).
27 + 73	
29 + 71	
31 + 69	
33 + 67	
35 + 65	**(1 + 1):** (100 = 3 + 97), (100 = 11 + 89), (100 = 17 + 83), (100 = 29 + 71) (100 = 47 + 53)
37 + 63	
39 + 61	
41 + 59	
43 + 67	
45 + 65	
47 + 53	
49 + 51	
----------	
51 + 49	
53 + 47	
↓	

As N continues to increase, the tendency is that [p, p+2], [p, p+4] and [p, p+6] will increase accordingly, their solution time is proportional to the N involved, each execution is independent. It seems we have ample empirical evidence to say that "Twin Primes" is correct, but it is impossible to exhaust all the numbers to have a conclusion.

[Study 2]: Since (1+1) and [p, p+2] both deal with odd prime numbers (favor rigor); a graphic method **(Appendix D)** in CHAPTER FOUR expressed clearly that there is no formula to resolve the "Twin Primes" --- but we know the [p, p+2] is "arithmetic" in nature and number theory (or arithmetic) is to study "the patterns of the numbers" and "elementary calculation technique".

From this perspective, the only polynomial procedure left to prove the validity of "Twin primes" is --- via an axiomatic method and reasoning.

4. THE SOLUTION

Let N = (5n ± 2), n = 1, 3, 5, 7, 9, 11, 13, 15, 17, 19...

Ascending order of odd numbers in the range of N= 100 below showed the distribution of odd numbers has (a particular form) to infinity; odd prime numbers exist in odd numbers and they can be easily identified via **(Appendix A-1)** in CHAPTER FOUR:

- 2	- 2	N	+ 2	+2
1	3	5	7	9
11	13	15	17	19
21	23	25	27	29
31	33	35	37	39
41	43	45	47	49
51	53	55	57	59
61	63	65	67	69
71	73	75	77	79
81	83	85	87	89
91	93	95	97	99
		↓		
		∞		

(a) Naturally, [p, p+2] only occurred in the vicinity of N because the last digit of odd primes can only be 1, 3, 7 or 9 (except prime 5).
(b) [p, p+2] will not appear at all the time due to nonlinearity of primes, but it will eventually reappear --- since there are infinitude of primes,
(c) The number pair of [p, p+2] is proportional to the N involved. Let: N = 40, 50 ...

N = 40: There are (3, 5), (5, 7), (11, 13), (17, 19), (29, 31)
N = 50: There are (3, 5), (5, 7), (11, 13), (17, 19), (29, 31), (41, 43)
N = 100: (3, 5), (5, 7), (11, 13), (17, 19), (29, 31), (41, 43), (59, 61), (71, 73)

The same polynomial procedure can be adapted to obtain every pair of [p, p+2] in the range of any N, however large.

External links are retrieved for general information of "Twin primes" only:

[1]: G.H Hardy; E. M. Wright (2008) [1938]. An introduction to the theory of numbers (rev. by D. R Heath-Brown and J. H Silverman,6th ed) Oxford University Press. ISBN 978-0-19-9211986-5 http:// google.com/book?id = rey9wfSaJ9EC&dq
[2]: Vinogradov.I.M. The method of trigonomitrical Sums in the theory of numbers. London Interscience, p 67. 1954
[3]: Hazewinkel, Michiel,ed (2001) " Congruence" Springer ISBN 978-1-55608-010-4
[4]: Weisstein, Eric W., "modular Arithmetic ", Math World
[5]: http://annals.math.princeton.edu/articles/7954
[6]: https:// simonsfoundatiin.org/quanta/20130519-unheralded-mathematcian-bridge-th...2/25/2014

EXHIBIT 9 A PROOF OF THE BEAL'S CONJECTURE (BC)

ABSTRACT

The 1993 Beal conjecture (BC) was accepted as an open problem in number theory because Andrew Beal"s (an amateur number theory enthusiast) attempt to resolve the elusive 1637 Fermat's Last Theorem (FLT) via common factor was consistent with the research activities of the 19th centuries. Moreover, Beal believed Fermat had a relatively simple non-geometry-based solution.

After resolved the Goldbach's conjecture and (FLT) elaborated in **(Exhibit A, B, C),** along with the (BDS) Ranking conjecture in **(Exhibit 3),** it became clear that the Beal's conjecture was correct, it can be resolved by a polynomial procedure, but the details are more complicated due to a comprehensive understanding structure of the whole numbers is a must.

1. THE PROBLEM

While Mr. Andrew Beal was investigating the generalization of the (FLT) in 1993, he formulated a conjecture that: Let A, B, C, x, y, z be positive integers with x, y, z > 2. If $A^x + B^y = C^z$, then A, B and C have a common factor.

2. EARLY HISTORY

Mathematicians have puzzled with: (a) the "pattern of the primes" and "concept of infinite", (b) dealt with questions of finding and describing the intersection of algebraic curves in early developments of algebraic geometry long before: the (450 BC) Pythagorean theorems, the Element (350 BC), the work of Sieve (270 BC) and Archimedes (250 BC).

In the 3rd century, "Arithmetica" introduced Diophantine geometry; it was a collection of problems giving numerical solutions of determinate and indeterminate equations. Diophantine studied integers, integers can be considered either in themselves or as solutions to equations. Diophantine was the first Greek mathematician who recognized fractions as numbers and allowed positive rational numbers for the coefficients and solutions. Diophantine equations are usually algebraic equations with integer coefficients, for which integer solutions are sought. Diophantine also studied rational points on curves and algebraic varieties. In other words, he showed how to obtain infinitely many of rational points satisfying a system of equations by giving a procedure that can be made into an algebraic expression (basic algebraic geometry in pure algebraic forms). Unfortunately, most of his work was lost or unexplained. Not much happened in number theory after the Greeks. Not much happened in mathematics after the Greeks.

In the 17th century, Galileo (1564-1642) was the father of "observational astronomy" and "modern science", a polymath in the field of mathematics, physics, engineering and natural philosophy.

In mathematics --- Galileo applied the standard passed down from ancient Greeks and Fibonacci (1170-1250); but superseded later by the algebraic methods of Descartes.

Rene' Descartes (1596-1630) introduced the analytic geometry primarily to study algebraic curves (reformulation of classic works on conic and cube). Using Descartes' approach, the geometric and logical arguments favored by the ancient Greeks for solving geometric problems could be replaced by doing algebra. Descartes also studied the tangent line problem. It was during the same period, Pierre de Fermat (1601- 1665) conjectured his own prime numbers and independently developed analytic geometry to study the properties of algebraic curves (those defined in Diophantine geometry), which is the manifestation of solutions of system of polynomial equations. Fermat also addressed Diophantine approximate equality to find maxima for functions and tangent lines to curves. However, most of Fermat's work was in margin note or private letters --- including his famous Fermat's last Theorem --- please revisit **(Exhibit C)** and this part of (FLT) was what Mr. Beal attempted to resolve:

"In about 1637, French mathematician Pierre de Fermat (1601- 1665) left a margin note in his copy of "Arithemtica" that this indeterminate Diophantine equation $A^N + B^N = C^N$ has no solution when N is an integer > 2"

3. OBSERVATION

(1) Proving the Goldbach's conjecture in **(Exhibit A and B)** illustrated that all the even number n > 2 can be expressed as the sum of two odd primes. For example: 4 = 2 + 2, 6 = 3 + 3, 8 = 3 + 5, 10 = 5 + 5, 10 = 3 + 7 ...... Positive integer 4, 6, 8, 10, 12 ... are composites with a common prime factor 2; and prime 2 is the building block of all the odd prime and composite numbers in the range of N = 40:

1 + 2 = $\underline{3}$	3 + 2 = $\underline{5}$	5 + 2 = $\underline{7}$	7 + 2 = 9	9 + 2 = $\underline{11}$	11 + 2 = $\underline{13}$
13 + 2 = 15.	15 + 2 = $\underline{17}$	17 + 2 = $\underline{19}$	19 + 2 = 21	21 + 2 = $\underline{23}$	23 + 2 = 25
25 + 2 = 27	27 + 2 = $\underline{29}$	29 + 2 = $\underline{31}$	31 + 2 = 33	33 + 2 = 35	35 + 2 = $\underline{37}$
37 + 2 = 39	so on to the forth				

Logically, infinitude of odd prime and composite numbers have the (prime 2) built-in; in other words, if a certain polynomial algorithm (of a particular form) can be found to hold the Beal's statement true with a common prime factor 2, then it should hold Beal's statement true for all the prime factors.

(2) **(Exhibit C), (Exhibit 3)** briefly mentioned "Common factor" and together with the **(Beal conjecture)** --- they were dealing with Diophantine equations. When we speak of Diophantine equations today, we are speaking of polynomial equations, in which integer solutions must be found. Polynomial is a mathematical expression that involves N's and N^2 s and N's raise to other powers

4. THE SOLUTION

The following [two studies] illustrated how to build extra structure to obtain two algorithms (of a particular form) that provided a complete system of solutions to hold the Beal's statement true infinitely via polynomial time procedures.

[Study 1]: $A^x + B^y = C^z$

Let: $A \neq B$ = positive integer, $x = y = 3$ (Beal required x, y, z > 2).

Hence: $A^x = A^3$

$B^y = (nA)^3$

$C^z = A^3 + (nA)^3 = A^3 (1 + n^3)$

Let: $A = (1 + n^3)$

Hence: $A^x = (1 + n^3)^3$

$B^y = [n (1 + n^3)]^3$

$C^z = (1 + n^3)^3 (1 + n^3) = (1 + n^3)^4$

$A^x + B^y = C^z$

Then: $(1 + n^3)^3 + [n (1 + n^3)]^3 = (1 + n^3)^4$ .... (1) Let: n = 1, 2, 3 ...

n = 1:	$2^3 + 2^3 = 2^4$	(8 + 8 = 16)	w/prime factor 2
n = 2:	$9^3 + 18^3 = 9^4$	(729 +5832 = 6561)	w/prime factor 3

$(1 + n^3)^3 + [n (1 + n^3)]^3 = (1 + n^3)^4$

n = 3: $(28)^3 + [3 (28)]^3 = (28)^4$ (21952 + 592704 = 614656)

n = 4: $65^3 + 260^3 = 65^4$ w/prime factor 5

n = 7: $343^3 + 2401^3 = 343^4$ w/prime factor 7

n = 8: $(1 + n^3)^3 + [n (1 + n^3)]^3 = (1 + n^3)^4$

$513^3 + [8 (513)]^3 = 513^4$

$513^3 + [4104]^3 = 513^4$ w/prime factor 3

n = 10: $(1 + n^3)^3 + [n (1 + n^3)]^3 = (1 + n^3)^4$

$(1001)^3 + [10 (1001)]^3 = (1001)^4$

$(1001)^3 + (10010)^3 = (1001)^4$ w/prime factor 7, 11

It is impossible to exhaust all the n(s); however, algorithm (1) is a polynomial equation (of a particular form) that its solution to one n can be adapted to solve all the other (n)s to hold the Beal's statement true infinitely when $A \neq B$.

[Study 2]: $A^x + B^y = C^z$

Let: $A = B$ = positive integer, $x \neq y$

Hence: $A^x = A^n$

$B^y = A^{n+1}$

$C^z = A^n + A^{n+1} = A^n (1 + A)$

Let n = 3, then: $A^x = A^3$ (Beal required x, y, z > 2).

$B^y = A^4$

$C^z = A^3 + A^4 = A^3 (1 + A)$

Let: $(1 + A) = n^3$

Variation: $A = n^3 - 1$

Hence: $A^x = A^3 = (n^3 - 1)^3$

$B^y = A^4 = (n^3 - 1)^4$

$C^z = A^3 (1 + A)$

$= (n^3 - 1)^3 (1 + n^3 - 1)$

$= (n^3 - 1)^{3\cdot} n^3$

$A^x + B^y = C^z$

Then: $(n^3 - 1)^3 + (n^3 - 1)^4 = (n^3 - 1) n^3$..... (2) Let: N = 2, 3, 4 ...

n = 2: $(8 - 1)^3 + (8 - 1)^4 = (8 - 1) (2)^3$

$7^3 + 7^4 = 14^3$ (343 +2401 = 2744) w/prime factor 7

n = 3: $26^3 + 26^4 = 78^3$ w/prime factor 2

n = 4: $63^3 + 63^4 = 252^3$ w/prime factor 3, 7

$(n^3 - 1)^3 + (n^3 - 1)^4 = (n^3 - 1) n^3$

n = 5: $(5^3 - 1)^3 + (5^3 - 1)^4 = (5^3 - 1) 5^3$

$(124)^3 + (124)^4 = (124) (5)^3$

1906624 + 236421376 = 238328000 w/ prime factor 2

$(n^3 - 1)^3 + (n^3 - 1)^4 = (n^3 - 1) n^3$

n = 10: $(999)^3 + (999)^4 = (999) (10)^3$

$(999)^3 + (999)^4 = (9990)^3$ w/ prime factor 3

It is impossible to exhaust all the n(s), but Algorithm (2) is a polynomial equation (of a particular form) that its solution to one n can be adapted to solve all the other (n)s to hold the Beal's statement true infinitely when A = B.

Algorithm (1) and (2) together provided a complete system of solution to hold the Beal's statement true infinitely.

5. THE CONCLUSION

(a) Beal's conjecture is correct that when A, B, C, x, y, z are positive integers with x, y, z > 2. If $A^x + B^y = C^z$, then A, B and C must have a common factor.

(b) But this proof of the Fermat's Last Theorem is not by (descent) and too long to be what Fermat had in minded.

External link is retrieved from:

[1]: A generalizations of FLT: The Beal conjecture and prize problem. Daniel Mauldin. Dec 1997 (notice of the AMS)

[1]:

EXHIBIT 10 TO CONSTRUCT (A 17-sided polygon)

ABSTRACT. Subject to you find the "Trisecting the Angle" leads to construct the "Regular Heptagon" elaborated in **(Exhibit D and E)** were credible, then the same procedure can be adapted to construct this 17-sided polygon showed in the next three pages should make sense; moreover, this polynomial procedure that can be adapted to construct a 5, 9, 11, 13, 15 or 19-sided polygon (in a reasonable sides).

1. THE PROBLEM

Mathematicians have puzzled with the paradoxes of the "pattern of the primes" and "concept of infinite" and how to construct a regular polygon of any given number of sides via ancient Greek "Straightedge and Compass construction".

2. EARLY HISOTRY

In the 17th century, French mathematician Pierre de Fermat (1601-1665) introduced his prime numbers as:

$F_n = 2^{2^n} + 1$ ....(A) when n = 0, 1, 2, 3, 4

n = 0: $F_0 = 3$

n = 1: $F_1 = 5$

n = 2: $F_2 = (16 + 1) = 17$

n = 3: $F_3 = (256 + 1) =257$

n = 4: $F_4 = (65536 +1) = 65537$.

And conjectured that all (infinity many) of the numbers in the form of (A) are primes.

In the 1730s, while Leonhard Euler (1707-1782) was investigating prime distribution, he became interested Fermat's work and proved in 1732(s) that when n = 5, $F_5 = 641 \times 6700417$ is composite. Therefore, Fermat's prime conjecture was flawed.

In the 18th century, Carl F. Gauss (1777-1855) proved in theory that a 17-sided polygon is constructible via ancient Greek "Straightedge and Compass construction" but no one was able to do so (including Gauss himself). Nevertheless, the math community accepted that Gauss proved the connection between the geometry and prime numbers.

Moreover, to honor Gauss's over all achievement in mathematics, a symbol of the "17-sided polygon" was place next to his grave at Gottingen, Germany.

3. THE SOLUTION

In 1992, professor Shi's high school student in Pairs France constructed the first (17-sided polygon) via "Straightedge and compass construction" after she understood the **(Exhibit D and E).** Moreover, the same polynomial procedure can be adapted to construct a regular polygon of any given number sides (5, 9, 11, 13, 15, 19 ...) not just prime numbers.

This 17-sided polygoan was constructed in 1991 by Mr. Shi's high school student in Paris, France --- based on "Strightedge and compass construction" --- the first time by human:

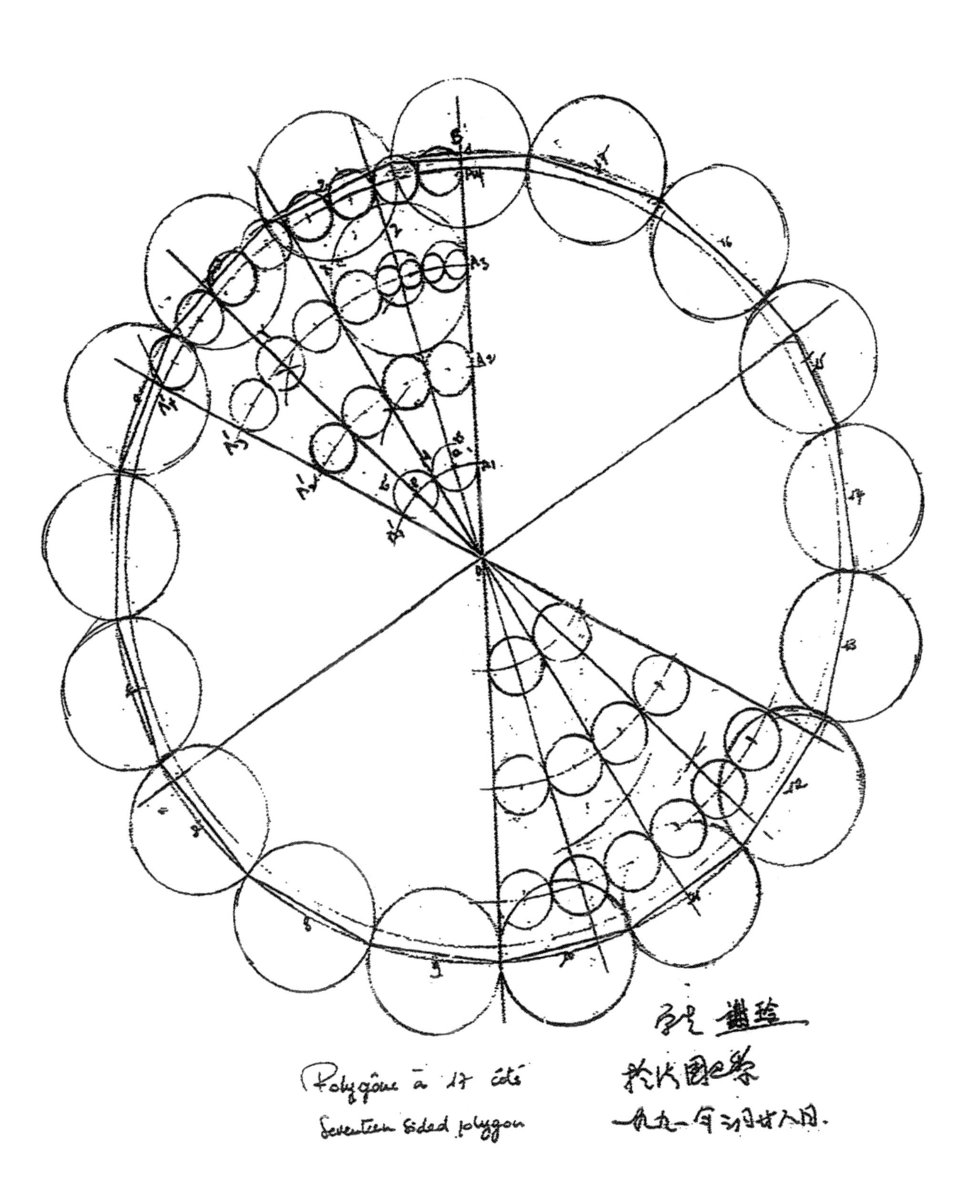

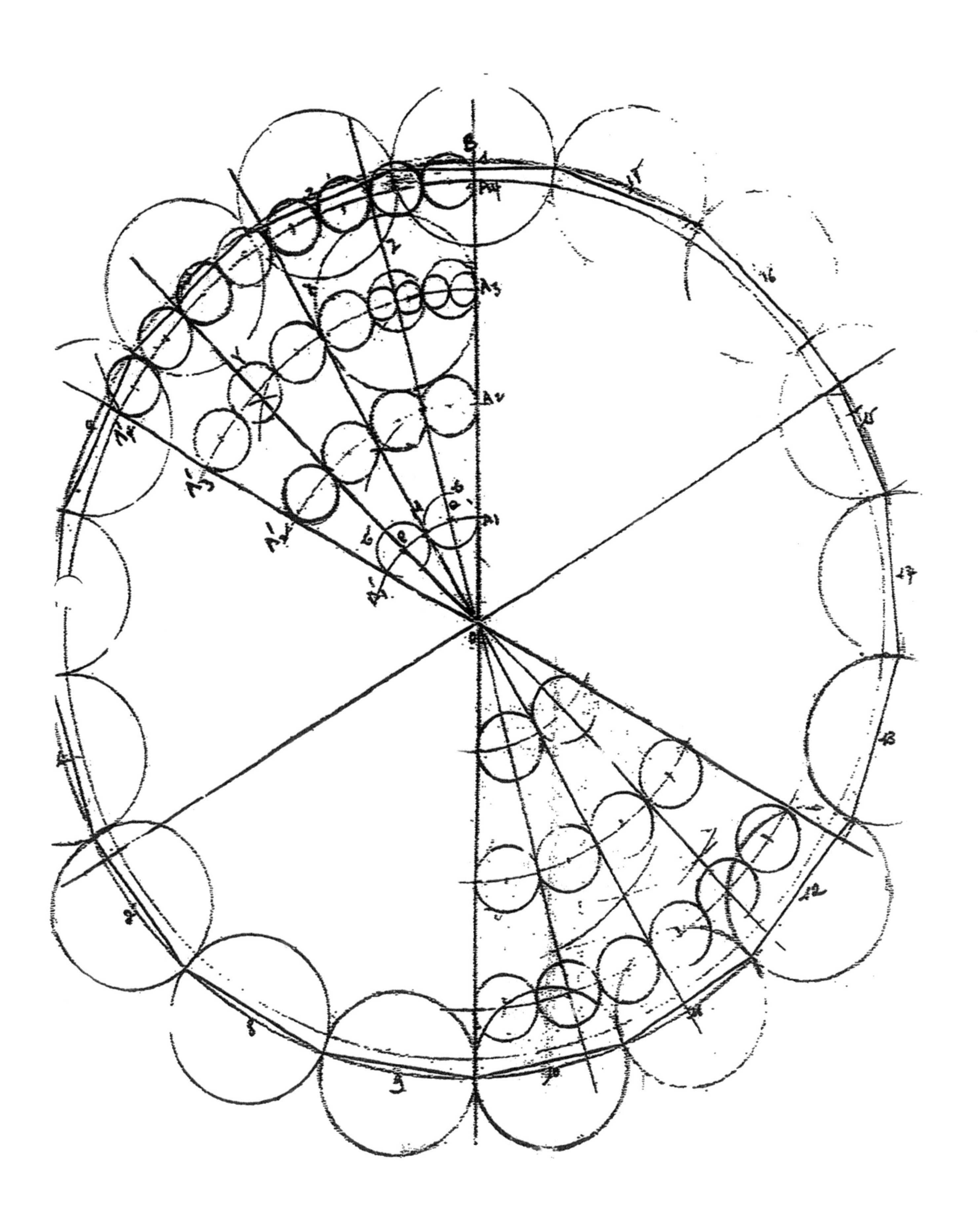

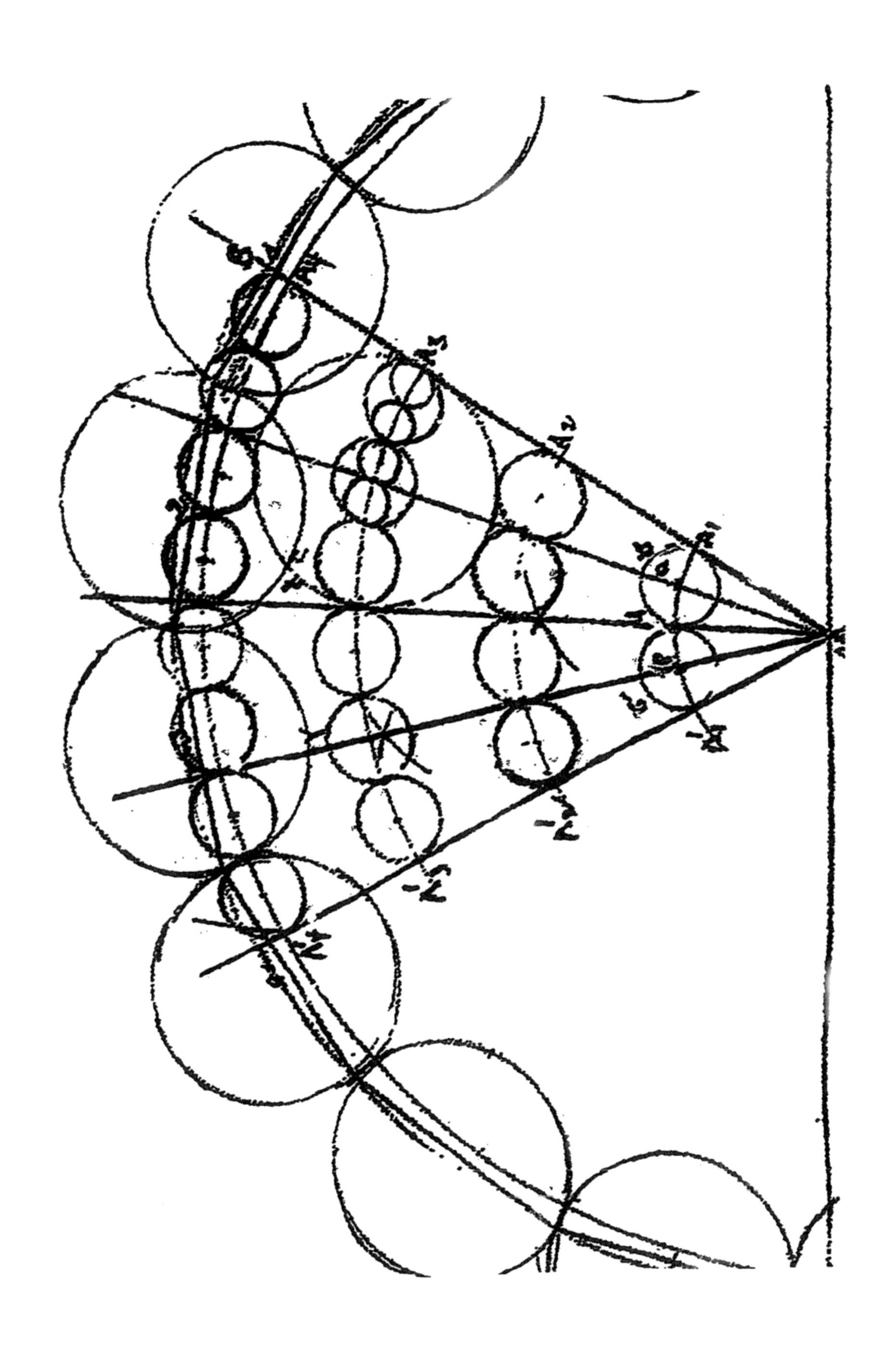

FOUR

Appendix

****
****
****

(Appendix A) Consists of letters of acknowledgement In the next 6 pages

The proclaimed proof of the "Goldbach's conjecture" was acknowledged by:

A 1989 letter from Chairman Arthur Wightman (Princeton math Department) to respond Mr. Lin's requests (a student of Mr. Shi) for a confidential verification.

The book "EXIST" was acknowledged by:

President and Mrs. George Bush in 1992, their small portrait was cut off from a Miami based Chinese newspaper, then pasted on by us

A 1994 letter from National Geographic Society mentioned about a color calculation chart that outperforms any computer and [...]

A 1994 letter from Vice President Al Gore

Who's Who in 1995

International Biographical Centre Cambridge CB2 3QP England in 1995

****
****
****

Princeton University Department of Mathematics
Fine Hall—Washington Road
Princeton, New Jersey 08544-1000

Mr. Dao-Zheng Lin
Flat D, 7/F Hooley Mansion
21-23 Wongneicheng Road
Hong Kong

Dear Mr. Lin,

Thank you for your letter of 28 May 1989, and the calculations it contained

I am very happy for you that your calculations have verified the validity of the Goldbach Conjecture in so many cases. May you have further success in your arithmetic explorations!

Sincerely,

Arthur Wightman

Arthur Wightman
Chairman

AW/cd

施先生

您的哥德巴赫猜想，我寄给Princeton大学的已有可喜的回复，下一步该怎么做，請指示，或您可直接与這间大学联络

致以热烈的祝賀！

道正敬上

SHI Feng SHeng

-Sept 17, 90

On behalf of the President

of the United States of America,

George Herbert Walker Bush,

the National Republican Congressional Committee

presents you with the

Official White House Portrait of

The President and Mrs. Bush,

Washington, D.C.

August, 1990

In grateful appreciation

美國總統布希及夫人予施馮生的贈照。

NATIONAL GEOGRAPHIC SOCIETY
1145 17th Street N.W., Washington, D.C. 20036-4688 U.S.A.

Senior Vice President

Dr. Feng Sheng Shi: April 12, 1994.

Dear Sir:

Your positive book of mathematics or knowledge has been called to our attention by our education Dept.

At times we publish books of this importance would you be so kind as to send A copy of your book "EXIST" to our Educational Science Division here in Washington, D. C. .

National Geographic Society 1145 17th Street N.W. Washington, D. C. 20036

sincerely

Michela England

M. E. / ab

This advanced book of mathematics was written by Dr. Fery Sheng Shi, the number one mathematician in the world today. It has been copywritten in the Library of Congress and also contains a chart that is only lines in color but can calculate faster than the computer or calculator. It is our belief that this chart can be used to advance the students in mathematics and also to sharpen their wits.

NATIONAL GEOGRAPHIC SOCIETY
145 17th Street N.W.

Postage OK
Southern Maryland Division 20790 - 9651

NATIONAL GEOGRAPHIC SOCIETY
1145 17th Street N.W.
Washington, D.C. 20036-4688

Vice President Al Gore

Dr. Feng Sheng Shi: April 1, 1994.

Dear Friend:

I have seen your book "EXIST" and I wish to personally congratulate on this wonderful book of mathematics and your chart. I feel very proud of you that you have decide to live among us, the land of the free where no man has to live in fear. I know you will continue your mathematics and I wish you every success. Your chart is very good to keep the young people quiet as they ride the plane they were very interested and I know it will improve their knowledge of mathematics.

If I can be of any help to you do not hesitate to call upon me.

Sincerely

Al Gore

Al Gore
Vice President of the United States

ice President Al Gore

Dr. Feng Sheng Shi

2114 Barritz #2

Miami Beach Florida.

By Authority of Its International Publications
The American Biographical Institute's
choice for

WHO'S WHO OF THE YEA

1995

is

Dr. Feng Sheng Shi

acknowledged for all the world and in appreciation of
outstanding accomplishments in this year of the Twentieth Century.

Issued and Authorized by the Institute's Directors

THIS PICTORIAL TESTIMONIAL OF ACHIEVEMENT AND DISTINCTION

proclaims throughout the world that

Dr Feng Sheng Shi

is the recipient of the above-mentioned Honour, granted by the Board of Editors of the

DICTIONARY OF INTERNATIONAL BIOGRAPHY

meeting in Cambridge, England, on the date set out below,

AND that the Board also resolves that a portrait photograph of

Dr Feng Sheng Shi

be attached to this Testimonial as verification of the Honour bestowed.

DICTIONARY OF INTERNATIONAL BIOGRAPHY
TWENTY FOURTH EDITION

Signed and sealed on the
30th October 1995

Authorized Officer C E Barker

(APPENDIX A-1) PATTERN OF THE ODD PRIMES (In a reasonable length)

In mathematics, an integer is any member of the set (… -2, -1, 0, +1, +2 …), including all the positive whole numbers, negative whole numbers and zero.

(a) Primes are any (infinitely many) positive integer ≥ 2 that can be divided by 1 and itself only (two divisors). Historically, there is no useful formula that yields all primes and no composites due to they are not polynomial. Polynomial is a mathematical expression that involves N's and N^2 s and N's raise to other powers.

Even number 4, 6, 8, 10 … have a common prime factor 2. Moreover, prime 2 is the building block of all the odd primes and composites:

1 + 2 = <u>3</u>	3 + 2 = <u>5</u>	5 + 2 = <u>7</u>	7 + 2 = 9	9 + 2 = <u>11</u>	11 + 2 = <u>13</u>
13 + 2 = 15	15 + 2 = <u>17</u>	17 + 2 = <u>19</u>	19 + 2 = 21	21 + 2 = <u>23</u>	so on to the forth

In essence, all the odd prime and composite numbers in the range of any N have a prime factor 2 (built in).

(b) (GC) deals with odd prime numbers; odd primes are any odd number ≥ 3 that can be <u>divided</u> by 1 and itself only (two divisors), so it is also true that --- the first three odd prime (3, 5, 7) multiplied with odd number 3, 5, 7, 9, 11, 13, 15 … respectively in the range of any N ≤ 5000 can be composite numbers only, what's left is a list of odd primes below:

3	5	7	11	13	17	19	23	29	31
37	41	43	47	53	59	61	67	71	73
79	83	89	97	101	103	107	109	113	127
131	137	139	149	151	157	163	167	173	179
181	191	193	197	199	211	223	227	229	233
239	241	251	257	263	269	271	277	281	283
293	307	311	313	317	331	337	347	349	353
359	367	373	379	383	389	397			

401	409	419	421	431	433	439	443	449	457
461	463	467	479	487	491	499	503	509	521
523	541	547	557	563	569	571	577	587	593
599	601	607	613	617	619	631	641	643	647
653	659	661	673	677	683	691	701	709	719
727	733	739	743	751	757	761	769	773	787
797	809	811	821	823	827	829	839	853	857
859	863	877	881	883	887	907	911	919	929
937	941	947	953	967	971	977	983	991	997

1009	1013	1019	1021	1031	1033	1039	1049	1051	1061
1063	1069	1087	1091	1093	1097	1103	1109	1117	1123
1129	1151	1153	1163	1171	1181	1187	1193	1201	1213
1217	1223	1229	1231	1237	1249	1259	1277	1279	1283
1289	1291	1297	1301	1303	1307	1319	1321	1327	1361
1367	1373	1381	1399	1409	1423	1427	1429	1433	1439

1447	1451	1453	1459	1471	1481	1483	1487	1489	1493
1499	1511	1523	1531	1543	1549	1553	1559	1567	1571
1579	1583	1597	1601	1607	1609	1613	1619	1621	1627
1637	1657	1663	1667	1669	1693	1697	1699	1709	1721
1723	1733	1741	1747	1753	1759	1777	1783	1787	1789
1801	1811	1823	1831	1847	1861	1867	1871	1873	1877
1879	1889	1901	1907	1913	1931	1933	1949	1951	1973
1979	1987	1993	1997	1999					

2003	2011	2017	2027	2029	2039	2053	2063	2069	2081
2083	2087	2089	2099	2111	2113	2129	2131	2137	2141
2143	2153	2161	2179	2203	2207	2213	2221	2237	2239
2243	2251	2267	2269	2273	2281	2287	2293	2297	2309
2311	2333	2339	2341	2347	2351	2357	2371	2377	2381
2383	2389	2393	2399	2411	2417	2423	2437	2441	2447
2459	2467	2473	2477	2503	2521	2331	2539	2543	2549
2551	2557	2579	2591	2593	2609	2621	2633	2647	2657
2659	2663	2671	2677	2683	2687	2689	2693	2699	2707
2711	2713	2719	2729	2731	2741	2749	2753	2767	2777
2789	2791	2797	2801	2803	2819	2833	2837	2843	2851
2857	2861	2879	2887	2897	2903	2909	2917	2927	2939
2953	2957	2963	2969	2971	2999				

3001	3011	3019	3023	3037	3041	3049	3061	3967	3079
3083	3089	3109	3119	3121	3137	3163	3167	3169	3181
3187	3191	3203	3209	3217	3221	3229	3251	3253	3257
3259	3271	3299	3301	3307	3313	3319	3323	3329	3331
3343	3347	3359	3361	3371	3373	3389	3391	3407	3413
3433	3449	3457	3461	3463	3467	3469	3491	3499	3511
3517	3527	3529	3533	3541	3547	3557	3559	3571	3581
3583	3593	3607	3613	3617	3623	3631	3637	3643	3659
3671	3673	3677	3691	3697	3701	3709	3719	3727	3733
3739	3761	3767	3769	3779	3793	3797	3803	3821	3823
3833	3847	3851	3853	3863	3877	3881	3889	3907	3911
3917	3919	3923	3929	3931	3943	3947	3967	3989	

4001	4003	4007	4013	4019	4021	4027	4049	4051	4057
4073	4079	4091	4093	4099	4111	4127	4129	4133	4139
4153	4157	4159	4177	4201	4211	4217	4219	4229	4231
4241	4243	4253	4259	4261	4271	4273	4283	4289	4297
4327	4337	4339	4349	4357	4363	4373	4391	4397	4409
4421	4423	4441	4447	4507	4513	4517	4519	4523	4547
4549	4561	4567	4583	4591	4597	4603	4621	4637	4639
4643	4649	4591	4657	4663	4673	4679	4691	4703	4721
4723	4729	4733	4751	4759	4783	4787	4789	4793	4799
4801	4813	4817	4831	4861	4871	4877	4889	4903	4909
4919	4931	4933	4937	4943	4951	4957	4967	4969	4973
4987	4993	4999							

In other words, we eliminated the composite numbers in the range of N = 5000.

In mathematics --- "Infinite" is greater in value than any specified numbers, however large, therefore, it is senseless to look for the largest primes. Nevertheless, this uncanny polynomial procedure can be adapted to explain where every odd primes will occur in N = 5500, 6000, 6500, 7000, 7500, 8000, or however large.

Appendix B

(Figure1)

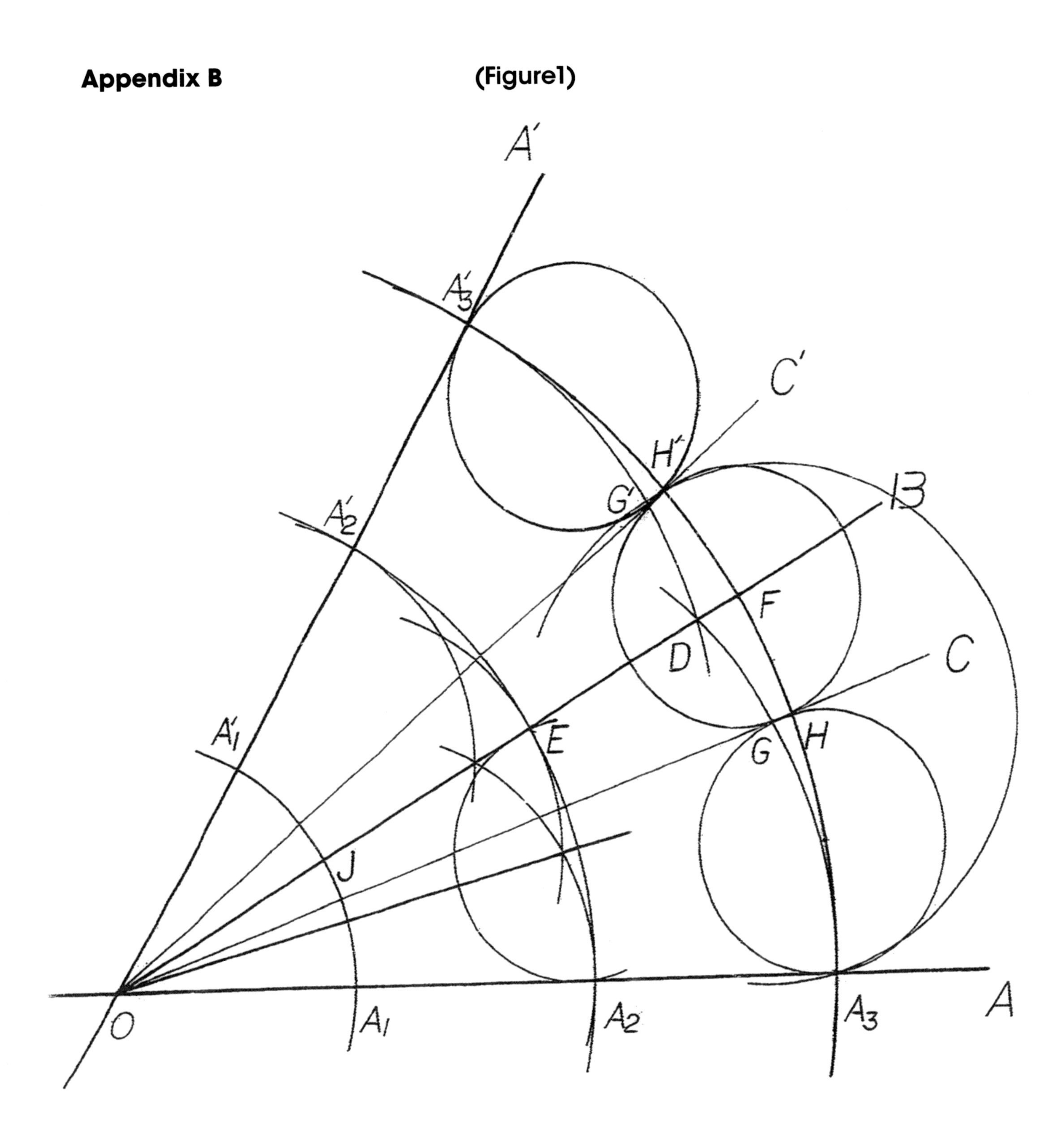

FIGURE 1 TRISECTING AN ANGLE

Appendix B-1 **(Heptagon)**

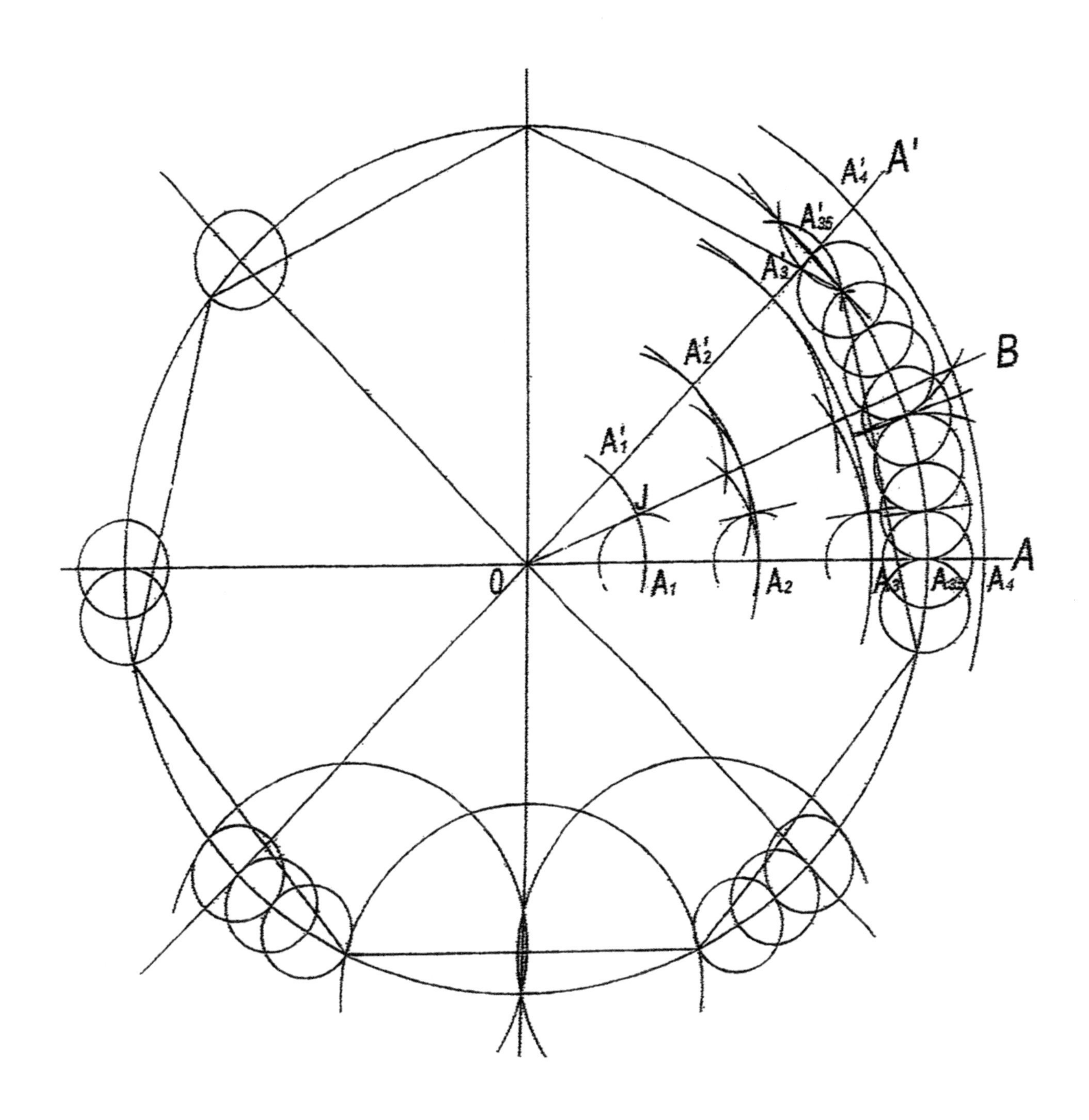

THE REGULAR HEPTAGON

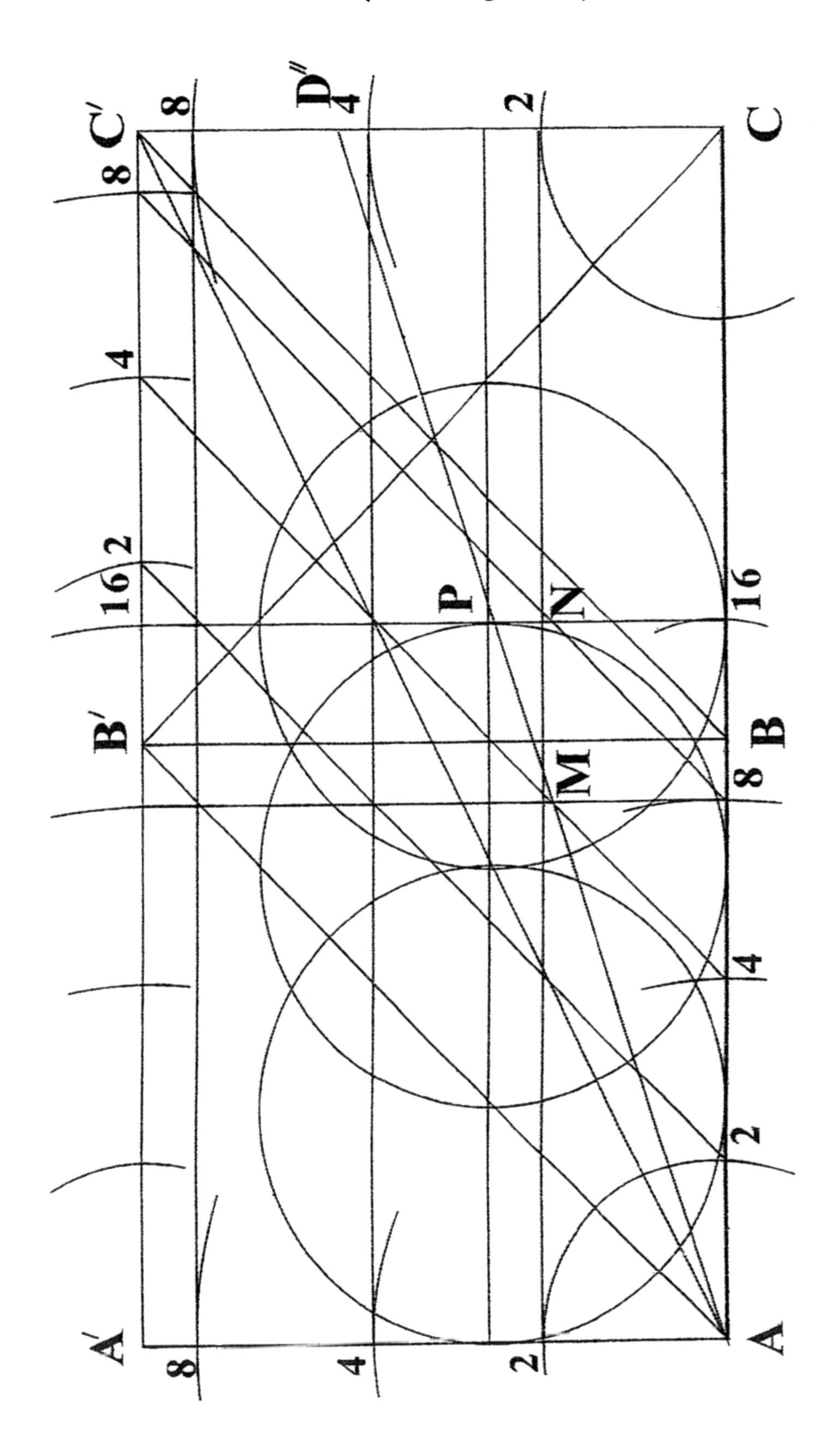

DOUBLING THE CUBE

Appendix C-1 **(Squaring the circle)**

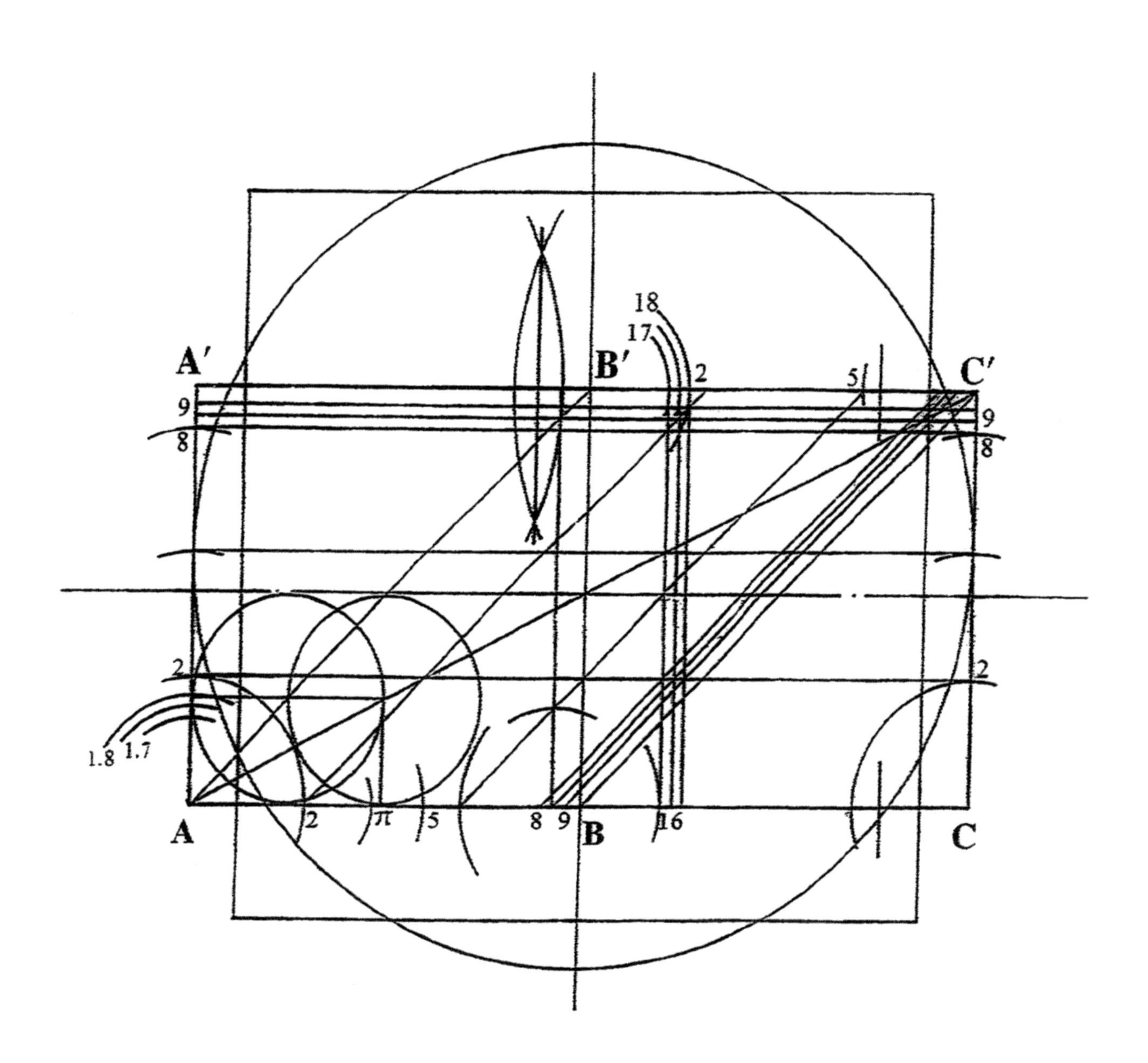

SOUARING THE CIRCLE

(Appendix D) A GRAPHIC PROOF OF THE GOLDBACH'S AND TWIN PRIMES CONJECTURE

==

Goldbach's conjecture and Twin Primes both deal with positive odd prime numbers; therefore, they are best understood together through a graphic study. Please use the color diagram in the next page:

==

(1) *Goldbach conjectured in about 1742 that: (a) all (infinity many) the even numbers ≥ 6 can be expressed as the sum of two odd primes, (b) any odd numbers ≥ 9 can be expressed as the sum of three odd numbers:*

This diagram consists of **green lines** (even number), **red lines** (prime numbers) and **blue lines** (odd numbers)

(a) In mathematics, an integer is any member of the set (... -2, -1, 0, +1, +2 ...), including all the positive whole numbers, negative numbers and zero.

(b) If integers encompass zero and negative numbers, then all even numbers are the sum of two primes. For example:

6 = 11 + (-5) 13 + (-7) 17 + (-11) ....
4 = 7 + (-3) 11 + (-7) 17 + (13) ....
2 = 5 + (-3) 7 + (-5) 13 + (-11) ....

Similarly, Zero is the sum of two prime numbers

(c) Naturally, all the odd numbers > 9 are the sum of three primes. Since 9 = 3 + N and N can be expressed as the sum of two primes. 9 will naturally be the sum of three primes. When N becomes larger, the odd numbers will be larger too.

(d) Moreover, this graphic proof coupled with proving the Goldbach's principle (1 + 1) enlightened to construct AN INFINITE CALCUALTION CHART elaborated in **(Exhibit B)** that outperforms any computer in calculation both in speed and range (however large or small).

(2) *Twin primes is defined as a pair of odd primes with a form of [p, p+2]. Since 2 is an even prime, therefore, the gap of (2, 3) is not considered as [p, p+2]. It asks to prove that there are infinitely many pairs of [p, p+2]*

(a) Noting green **lines** (even number) parallel to each other infinitely.

(b) **Red lines** (odd prime numbers) exist in **blue lines** (odd numbers) only, and they parallel to each other to infinity; therefore, there is no meaning structure to formulate any kind of formula to speak of; but we know that [p p+2] is a mathematical true - since there are infinitude of prime numbers.

(Appendix D-1)

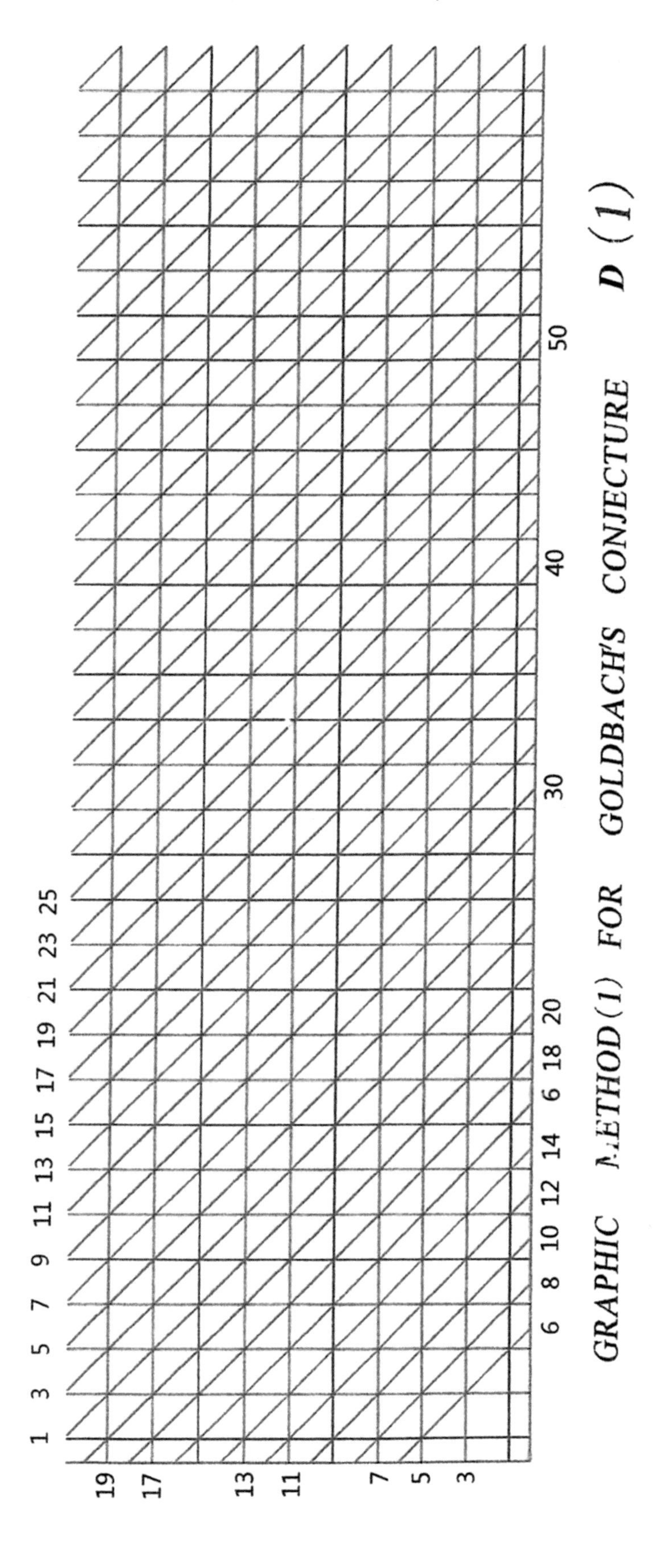
1 3 5 7 9 11 13 15 17 19 21 23 25
6 8 10 12 14 16 18 20 30 40 50
19 17 13 11 7 5 3
GRAPHIC METHOD(1) FOR GOLDBACH'S CONJECTURE D(1)

Appendix E-1

and

Appendix E-2

For your convenience:

The following two large color chart E-1 and E-2 can be viewed or retrieved from our website

PrimeDistributionRevealed.com
(Password 51760)

Appendix E-1 and Appendix E-2

Appendix E-1

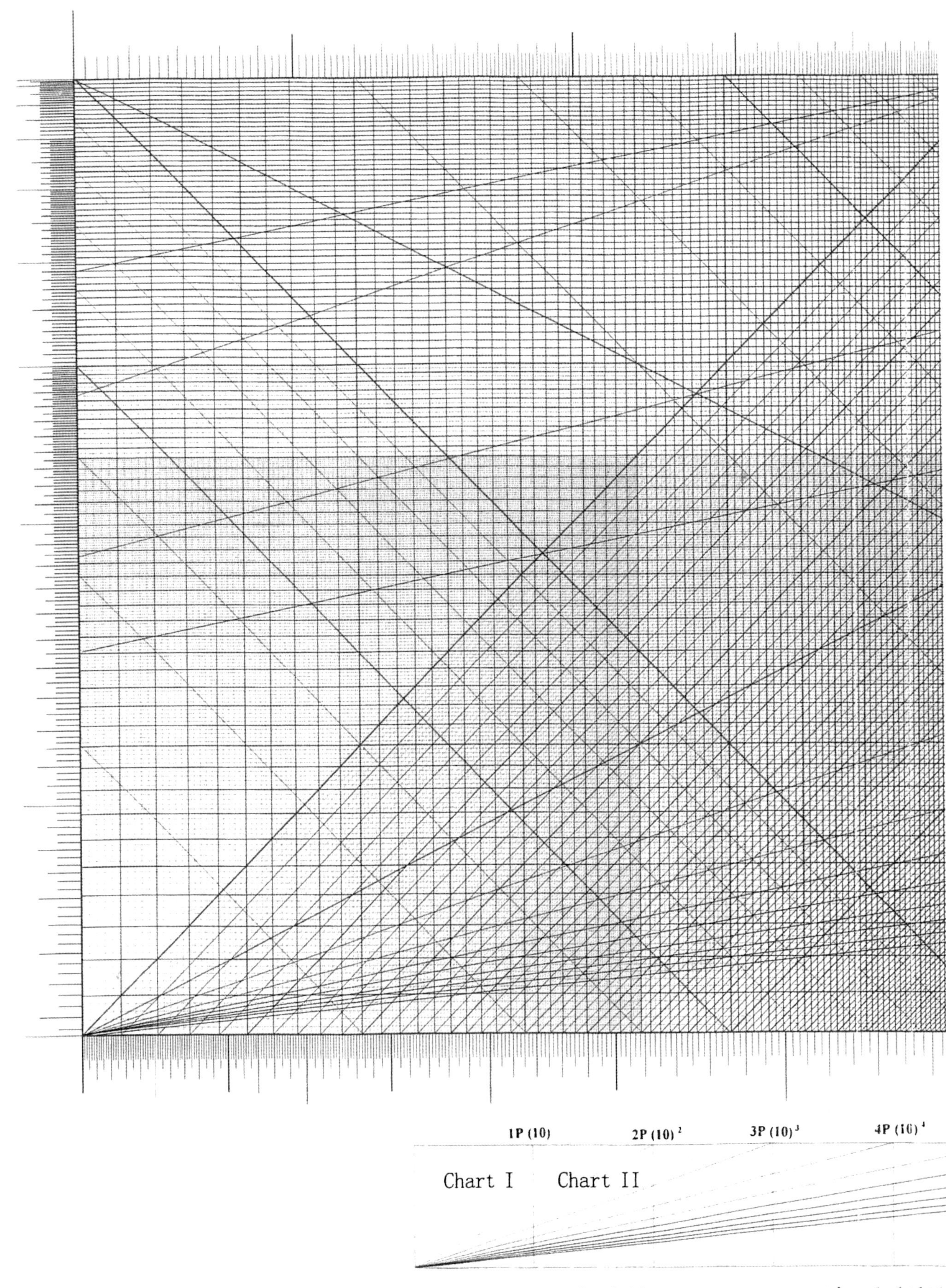

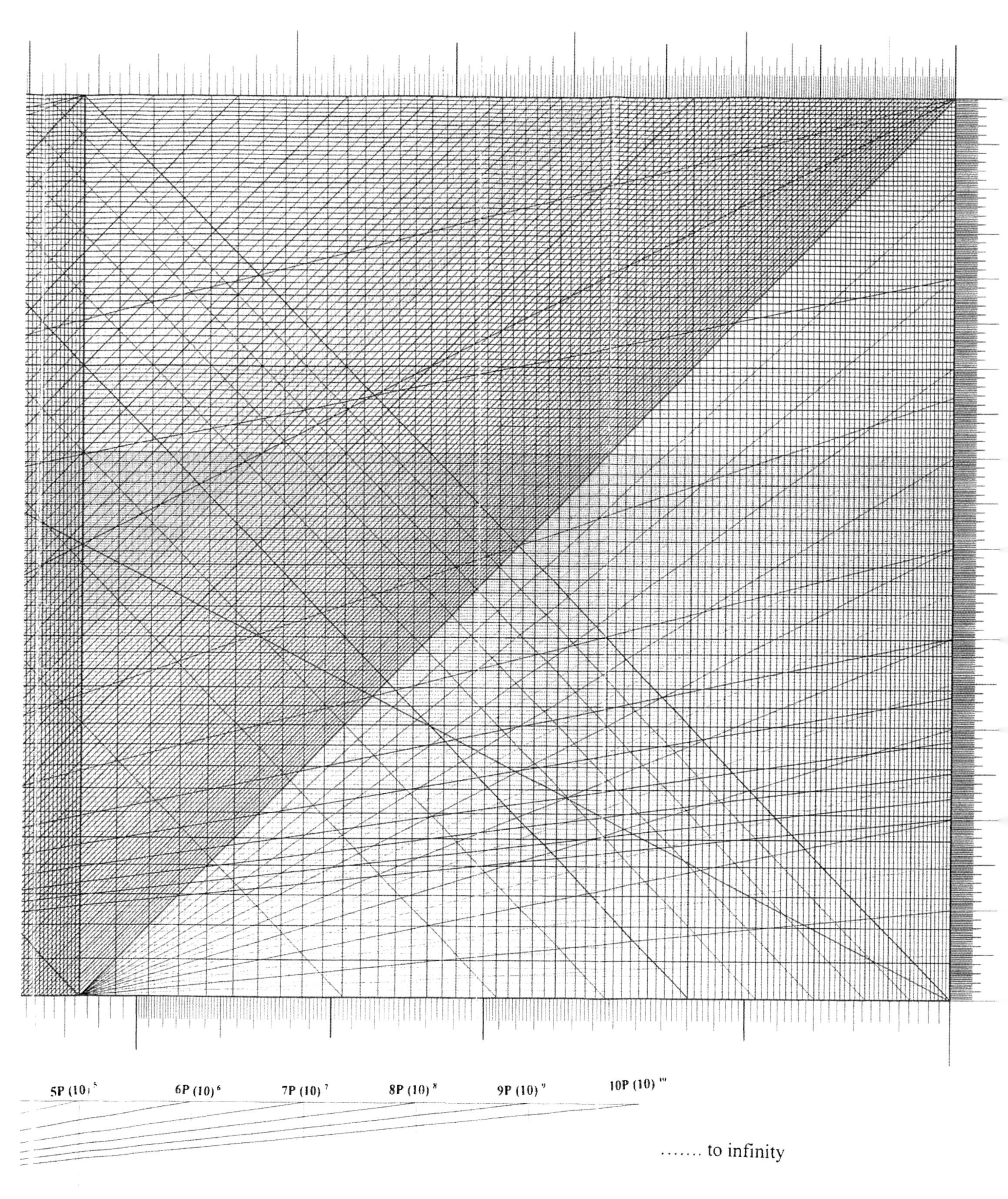

+1+1+1+1+1+1+1

Appendix E2

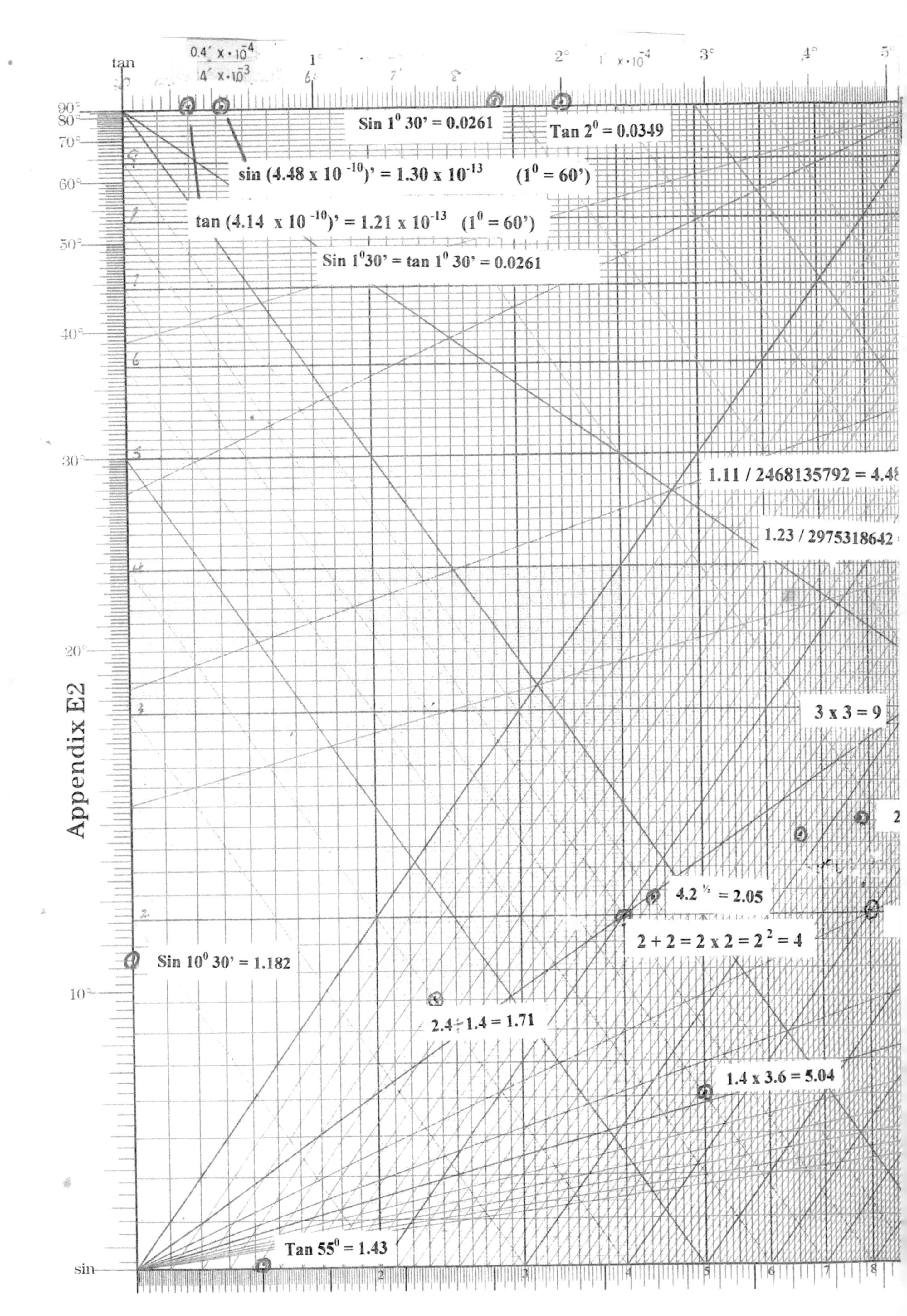

tan
sin
90°
80°
70°
60°
50°
40°
30°
20°
10°
1°
2°
3°
4°
5°
0.4' x·10^-4
4' x·10^-3
Sin 1^0 30' = 0.0261
Tan 2^0 = 0.0349
sin (4.48 x 10^-10)' = 1.30 x 10^-13 (1^0 = 60')
tan (4.14 x 10^-10)' = 1.21 x 10^-13 (1^0 = 60')
Sin 1^0 30' = tan 1^0 30' = 0.0261
1.11 / 2468135792 = 4.48
1.23 / 2975318642
3 x 3 = 9
4.2 ½ = 2.05
2 + 2 = 2 x 2 = 2^2 = 4
Sin 10^0 30' = 1.182
2.4 ÷ 1.4 = 1.71
1.4 x 3.6 = 5.04
Tan 55^0 = 1.43

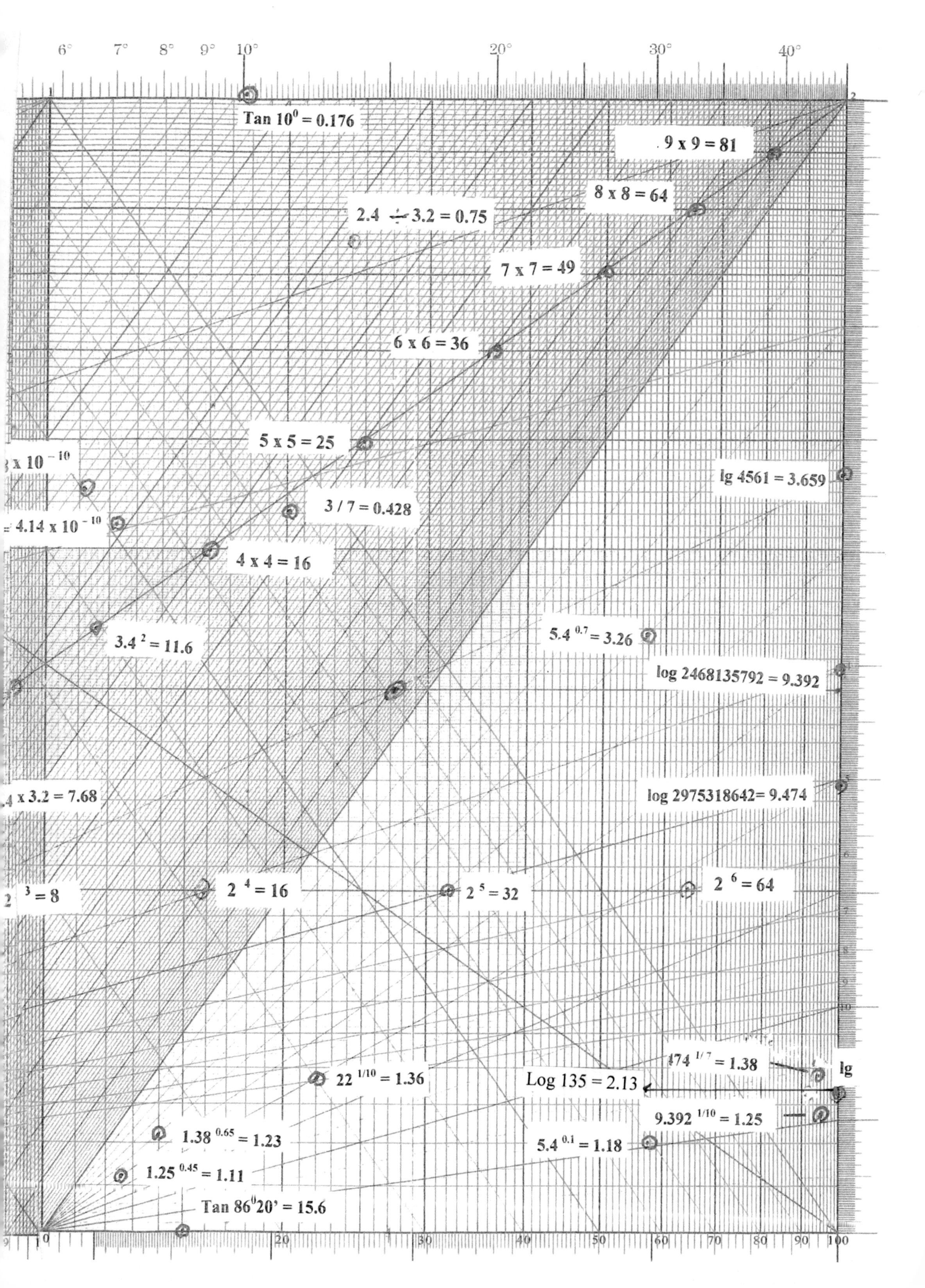
6°
7°
8°
9°
10°
20°
30°
40°
Tan 10⁰ = 0.176
9 x 9 = 81
8 x 8 = 64
2.4 ÷ 3.2 = 0.75
7 x 7 = 49
6 x 6 = 36
5 x 5 = 25
x 10 ⁻¹⁰
lg 4561 = 3.659
3 / 7 = 0.428
4.14 x 10 ⁻¹⁰
4 x 4 = 16
5.4 ⁰·⁷ = 3.26
3.4 ² = 11.6
log 2468135792 = 9.392
x 3.2 = 7.68
log 2975318642= 9.474
2 ³ = 8
2 ⁴ = 16
2 ⁵ = 32
2 ⁶ = 64
474 ¹/⁷ = 1.38
lg
22 ¹/¹⁰ = 1.36
Log 135 = 2.13
9.392 ¹/¹⁰ = 1.25
1.38 ⁰·⁶⁵ = 1.23
5.4 ⁰·¹ = 1.18
1.25 ⁰·⁴⁵ = 1.11
Tan 86⁰20' = 15.6
10
20
30
40
50
60
70
80
90
100

Printed in the United States
by Baker & Taylor Publisher Services